Springer Theses

Recognizing Outstanding Ph.D. Research

Aims and Scope

The series "Springer Theses" brings together a selection of the very best Ph.D. theses from around the world and across the physical sciences. Nominated and endorsed by two recognized specialists, each published volume has been selected for its scientific excellence and the high impact of its contents for the pertinent field of research. For greater accessibility to non-specialists, the published versions include an extended introduction, as well as a foreword by the student's supervisor explaining the special relevance of the work for the field. As a whole, the series will provide a valuable resource both for newcomers to the research fields described, and for other scientists seeking detailed background information on special questions. Finally, it provides an accredited documentation of the valuable contributions made by today's younger generation of scientists.

Theses are accepted into the series by invited nomination only and must fulfill all of the following criteria

- They must be written in good English.
- The topic should fall within the confines of Chemistry, Physics, Earth Sciences, Engineering and related interdisciplinary fields such as Materials, Nanoscience, Chemical Engineering, Complex Systems and Biophysics.
- The work reported in the thesis must represent a significant scientific advance.
- If the thesis includes previously published material, permission to reproduce this must be gained from the respective copyright holder.
- They must have been examined and passed during the 12 months prior to nomination.
- Each thesis should include a foreword by the supervisor outlining the significance of its content.
- The theses should have a clearly defined structure including an introduction accessible to scientists not expert in that particular field.

More information about this series at http://www.springer.com/series/8790

Björn Richerzhagen

Mechanism Transitions in Publish/Subscribe Systems

Adaptive Event Brokering for Location-based Mobile Social Applications

Doctoral Thesis accepted by
the Technische Universität Darmstadt, Darmstadt, Germany

Author
Dr. Björn Richerzhagen
Multimedia Communications Lab (KOM),
Department of Electrical Engineering
and Information Technology
Technische Universität Darmstadt
Darmstadt, Germany

Supervisor
Prof. Ralf Steinmetz
Multimedia Communications Lab (KOM),
Department of Electrical Engineering
and Information Technology
Technische Universität Darmstadt
Darmstadt, Germany

ISSN 2190-5053 ISSN 2190-5061 (electronic)
Springer Theses
ISBN 978-3-030-06463-1 ISBN 978-3-319-92570-7 (eBook)
https://doi.org/10.1007/978-3-319-92570-7

Printed on acid-free paper

This Springer imprint is published by the registered company Springer Nature Switzerland AG
The registered company address is: Gewerbestrasse 11, 6330 Cham, Switzerland

Supervisor's Foreword

It is a great pleasure to introduce Dr. Björn Richerzhagen's thesis, accepted for publication within Springer Theses and awarded with a prize for his outstanding work. Dr. Richerzhagen joined our Multimedia Communications Lab at Technische Universität Darmstadt in January 2013. He conducted his research within a full position as Research Assistant in the Collaborative Research Centre (CRC) 1053 MAKI—Multi-Mechanisms Adaptation for the Future Internet—approved by the German Research Foundation (DFG). He completed it with an oral defense in July 2017.

In this thesis, the new concept of transitions of mechanisms in communication and distributed systems is outlined. Transitions allow to conceive and build extremely scalable systems. For the first time—worldwide—transitions are studied in the context of publish/subscribe systems for event-based communication. Being the foundation of an increasing number of location-based mobile social applications, such systems need to be adapted to highly dynamic load patterns and to diverse and varying environmental conditions. As no single publish/subscribe system can cope with the extreme requirements resulting from human behavior and complex heterogeneous communication systems, Björn Richerzhagen proposes to execute transitions at runtime to adapt the overall communication system. For the first time in research, results are presented covering: (i) the design of transition-enabled publish/subscribe mechanisms for location-based filtering and locality-aware dissemination of events; (ii) the seamless execution of transitions in a mobile environment; and (iii) understanding the impact of transitions on the performance and cost of the communication system in a highly relevant and complex application scenario. In addition to the outstanding contributions regarding the seamless execution of transitions, the thesis constitutes a significant scientific advancement for transition-enabled communication systems as a necessary foundation for the future Internet.

Darmstadt, Germany
February 2018

Prof. Dr.-Ing. Ralf Steinmetz

Parts of this thesis have been published in the following documents:

Alexander Frömmgen, Björn Richerzhagen, Julius Rückert, David Hausheer, Ralf Steinmetz, and Alejandro Buchmann. "Towards the Description and Execution of Transitions in Networked Systems." In: *Proc. 9th International Conference on Autonomous Infrastructure, Management and Security (AIMS)*. IFIP. June 2015, pp. 17–29.

Börn Richerzhagen and Ralf Steinmetz. "Towards an Adaptive Publish/Subscribe Approach Supporting Transitions." In: *Proc. 7th International Conference on Autonomous Infrastructure, Management and Security (AIMS), Ph. D. Workshop*. IFIP. June 2013, pp. 84–87.

Börn Richerzhagen, Dominik Stingl, Ronny Hans, Christian Groß, and Ralf Steinmetz. "Bypassing the Cloud: Peer-assisted Event Dissemination for Augmented Reality Games." In: *Proc. 14th IEEE Conference on Peer-to-Peer Computing (P2P)*. Sept. 2014, pp. 1–10.

Börn Richerzhagen, Stefan Wilk, Julius Rückert, Denny Stohr, and Wolfgang Effelsberg. "Transitions in Live Video Streaming Services." In: *Proc. Workshop on Design, Quality and Deployment of Adaptive Video Streaming (VideoNEXT)*. ACM. 2014, pp. 37–38.

Börn Richerzhagen, Dominik Stingl, Julius Rückert, and Ralf Steinmetz. "Simonstrator: Simulation and Prototyping Platform for Distributed Mobile Applications." In: *Proc. 8th International Conference on Simulation Tools and Techniques (SIMUTOOLS)*. ACM. Aug. 2015, pp. 99–108.

Börn Richerzhagen, Marc Schiller, Max Lehn, Denis Lapiner, and Ralf Steinmetz. "Transition-enabled Event Dissemination for Pervasive Mobile Multiplayer Games." In: *Proc. 16th International Symposium on a World of Wireless, Mobile and Multimedia Networks (WoWMoM)*. IEEE. 2015.

Börn Richerzhagen, Alexander Wagener, Nils Richerzhagen, Rhaban Hark, and Ralf Steinmetz. "A Framework for Publish/Subscribe Protocol Transitions in Mobile Crowds." *In: Proc. 10th International Conference on Autonomous Infrastructure, Management and Security (AIMS)*. IFIP. June 2016, pp. 1–14.

Börn Richerzhagen, Nils Richerzhagen, Julian Zobel, Sophie Schönherr, Boris Koldehofe, and Ralf Steinmetz. "Demo: Seamless Transitions Between Filter Schemes for Location-based Mobile Applications." In: *Demonstrations of the 41st IEEE Conference on Local Computer Networks (LCN-Demos)*. Nov. 2016, pp. 1–3.

Börn Richerzhagen, Nils Richerzhagen, Julian Zobel, Sophie Schönherr, Boris Koldehofe, and Ralf Steinmetz. "Seamless Transitions Between Filter Schemes for Location-based Mobile Applications." In: *Proc. 41st IEEE Conference on Local Computer Networks (LCN)*. Nov. 2016, pp. 1–9.

Björn Richerzhagen, Nils Richerzhagen, Sophie Schönherr, Rhaban Hark, and Ralf Steinmetz. "Stateless Gateways—Reducing Cellular Traffic for Event Distribution in Mobile Social Applications." In: *Proc. 25th International Conference on Computer Communication and Networks (ICCCN)*. IEEE. Aug. 2016, pp. 1–9.

Nils Richerzhagen, Björn Richerzhagen, Dominik Stingl, and Ralf Steinmetz. "The Human Factor: A Simulation Environment for Networked Mobile Social Applications." In: *Proc. International Conference on Networked Systems (NetSys)*. IEEE. Mar. 2017, pp. 1–8.

Matthias Wichtlhuber, Björn Richerzhagen, Julius Rückert, and David Hausheer. "TRANSIT: Supporting Transitions in Peer-to-Peer Live Video Streaming." In: *Proc. IFIP Networking Conference (IFIP Networking)*. IEEE. 2014, pp. 1–9.

Contents

Acronyms

AoI	Area of Interest
API	Application Programming Interface
AS	Autonomous System
CEP	Complex Event Processing
CTS	Clear to Send
DHT	Distributed Hash Table
DTN	Delay-tolerant Networking
GPS	Global Positioning System
IGMP	Internet Group Management Protocol
IP	Internet Protocol
ISO	Organization for Standardization
LTE	Long-Term Evolution
MAC	Medium Access Control
MANET	Mobile Ad Hoc Network
MBR	Minimum Bounding Rectangle
MLD	Multicast Listener Discovery
NAT	Network Address Translation
NVE	Networked Virtual Environment
OS	Operating System
OSI	Open Systems Interconnection
OSM	OpenStreetMap
P2P	Peer-to-Peer
PDU	Protocol Data Unit
PoI	Point of Interest
QoE	Quality of Experience
QoS	Quality of Service
RTS	Request to Send
SCTP	Stream Control Transmission Protocol
SDN	Software-defined Network
SDWN	Software-defined Wireless Network

Chapter 1
Introduction

The expansion of mobile broadband networks and the advance of smartphones into our every-day life have changed the way we interact with communication technology. With 3.7 billion mobile broadband subscriptions in 2016 [17], ubiquitous access to the Internet is taken for granted. By 2020, about 30% of all Internet traffic is predicted to be caused by smartphones according to Cisco [11], for the first time exceeding the amount of traffic originating from PCs. The highest growth with approximately 24% each year until 2020 is predicted for *location-based services* [12].

Location-based services exploit the current location of the mobile device to determine the relevance of information for a user [60]. Prominent examples are navigation applications like Google Maps, recommender portals like Yelp or TripAdvisor (*top five restaurants nearby?*), and location-based advertising [15, 70].

At the same time, there is a shift towards *mobile social applications*, comparable to the success of online social networks. A mobile social application enables and fosters direct interaction between users. Such interaction already follows coarse location-dependent patterns in basic messaging applications such as *WhatsApp* or *Telegram* [59, 67]. However, locality in the interaction between users becomes even more significant for mobile social applications that explicitly incorporate location-based services.

Examples for the resulting *location-based mobile social applications* constitute communities for sports and hobbies (e.g., *Sporty*, *PlayWith*, or *bvddy*), location-based messaging applications (e.g., *Lokin*, *happn*) and mobile augmented reality games such as Google's *Ingress* or the recently introduced *Pokémon Go*. In augmented reality games, the physical world surrounding a user is augmented with game-specific information and interaction possibilities. Thus, users need to move around in the physical world and interact with real-world objects and other users in their proximity [16]. In 2015, the traffic caused by augmented reality applications increased by a factor of four compared to 2014 [12]. Based on the tremendous success of Ingress and Pokémon Go, the 61-fold increase predicted by Cisco until 2020 seems reasonable.

B. Richerzhagen, *Mechanism Transitions in Publish/Subscribe Systems*,
Springer Theses, https://doi.org/10.1007/978-3-319-92570-7_1

This traffic growth and the increasing significance of direct interaction among users in location-based mobile social applications pose major challenges to a communication system. Efficient dissemination of content created and consumed by mobile users requires mechanisms for location-based filtering and, consequently, locality-aware dissemination. These mechanisms need to adapt to the inherent dynamics in terms of mobility and network heterogeneity of mobile social applications.

In this thesis, we propose a methodology for and, consequently, a realization of, transitions between publish/subscribe mechanisms to address the aforementioned challenges. In the following, we motivate our approach to adaptive event brokering for location-based mobile social applications.

1.1 Motivation for Adaptive Event Brokering

The *publish/subscribe* paradigm [19] is particularly well suited to model direct interaction between users and the interest-based communication pattern of mobile social applications. Users specify their interest in certain contents (hereafter referred to as *events*) through *subscriptions* that are issued to a *broker*. The broker (or, for scalability reasons, a network of brokers) filters incoming events based on these subscriptions and determines the set of users to which the events need to be disseminated. In this thesis, we focus on the adaptive combination of location-based filtering and locality-aware dissemination of events as core components of event brokering.

Early research on publish/subscribe systems focused on the organization of distributed broker networks to increase scalability or to account for potentially unreliable brokers [7, 37, 43, 46]. Techniques for efficient bundling [63] and merging of similar subscriptions [25, 41, 45] were proposed to reduce the load caused by forwarding events and subscriptions in the broker network. While these approaches initially focused on static clients, broker networks were later extended to support mobile publishers and subscribers. Here, research focused on dealing with connection failures [5, 42] and the migration of client state between brokers [9, 10, 20, 24, 30, 44]. Bandwidth limitations of the cellular connection were addressed by aggregating events before sending them to mobile clients [31, 61] or even by discarding certain events [42]. While the aforementioned approaches support physical mobility of clients, they do not enable location-based filtering of events, i. e., logical mobility.

Support for location-based filtering of events can either be added by extending existing publish/subscribe systems [4, 13, 20, 24, 33, 69, 69] or by introducing new subscription models for location-based filters [1, 3, 8, 18, 38, 68]. By extending existing systems, approaches can benefit from optimizations within channel- or attribute-based publish/subscribe systems [19]. However, they lack support for frequent location updates or fine-grained location-based subscriptions [13, 20]. Therefore, other approaches introduce new subscription models to offer location-based filtering as a native operation in the publish/subscribe system. The resulting dynamic nature of subscriptions poses a challenge to optimizations for broker networks [6], such as subscription merging [25, 45] and interest clustering [34–36].

While the aforementioned approaches support location-based filtering, they do not exploit the locality properties within the broker network. Locality of interaction is a key property of recent location-based mobile social applications such as the aforementioned augmented reality games, where user interaction is mostly determined by physical proximity. Several mechanisms exist to support locality-aware event dissemination, comprising (i) offloading mechanisms [47, 55], where predetermined gateways communicate with the broker on behalf of other users, and (ii) fully decentralized brokering within ad hoc networks formed by mobile users [21, 32, 39, 40]. By exploiting locality and mobile devices' distinct communication interfaces, these mechanisms reduce the load on the cellular infrastructure.

The efficiency of the aforementioned mechanisms depends on environmental conditions, such as the number of users, their mobility characteristics, and the availability of additional infrastructure (e. g., Wi-Fi access points). The dynamics of these conditions and the frequency of changes varies substantially for location-based mobile social applications [2]. Thus, the underlying publish/subscribe system needs to adapt accordingly. This is currently achieved by (i) reconfiguring the mechanism itself [26, 28, 29] or (ii) switching to another mechanism in a middleware [14, 62].

While the former can be achieved at low cost and high frequency, the range of conditions covered by reconfiguring a single mechanism is limited [2]. The latter approach can cover a wider range of environmental conditions; however, exchanging mechanisms is a costly and slow operation in current publish/subscribe middleware platforms, preventing its application in highly dynamic scenarios [14]. Still, according to [2], systems need to "...self-adapt autonomously by dynamically combining most suitable data distribution methods and techniques ...". Recently, coordinated mechanism transitions [23, 48, 66] beyond the limitations of middleware platforms have proven to enable systems to adapt to a wide range of conditions at low cost [56, 58, 65]. Therefore, (i) composing mechanisms for location-based filtering and locality-aware event dissemination and (ii) realizing transitions between different combinations of these mechanisms constitute essential steps towards adaptive event brokering for mobile social applications, as addressed in this thesis.

1.2 Research Challenges

Mobile social applications impose challenges on mechanisms for location-based filtering and locality-aware dissemination of events. The following research challenges affect our proposed methodology for transitions between these mechanisms and the realization of the respective transitions in a publish/subscribe system.

Challenge: *Adaptation to user mobility and interaction patterns*

In location-based mobile social applications, user mobility and interaction with the application are closely intertwined. Application-specific interaction patterns, e. g., with points of interest, affect the efficiency of individual publish/subscribe mechanisms due to the resulting heterogeneity in terms of workload and client mobility.

Addressing the resulting workload and mobility characteristics in a publish/subscribe system pose significant challenges given the limitations of individual mechanisms for location-based filtering and locality-aware dissemination of events. Consequently, our methodology for transitions between distinct mechanisms is required to adapt the publish/subscribe system to these application-specific characteristics.

Challenge: *Efficient utilization of heterogeneous networks*

Cellular plans are limited in terms of bandwidth and traffic volume, requiring efficient utilization of available resources to avoid decreasing user satisfaction [27]. Consequently, locality-aware event dissemination mechanisms can utilize Mobile Ad Hoc Networks MANET to offload traffic from the cellular network, relying on distinct communication interfaces available on today's mobile devices. Here, communication characteristics change rapidly as a consequence of user mobility. Therefore, the respective mechanisms need to deal with sudden degradations in bandwidth and even complete loss of connectivity between devices. Additionally, the capacity of a MANET is limited, requiring careful consideration of how and when ad hoc communication is utilized. Finding a suitable combination of centralized (cellular) communication and decentralized (ad hoc) communication and adapting the choice of mechanisms to address mobility while still offloading the cellular network pose significant challenges. This includes the selection of suitable gateways, i. e., users that receive an event via the cellular connection and forward it in an ad hoc fashion.

1.3 Research Goals and Contributions

The main goal of this work is the development of *a methodology for transitions between publish/subscribe mechanisms* to provide adaptive event brokering for location-based mobile social applications, and, consequently, *their realization and evaluation*. This objective is divided into the following three major research goals.

Research Goal 1: *Identification of mechanisms and potential transitions*

To identify promising mechanisms and potential candidates for transitions, a classification of existing approaches to adaptive event brokering needs to be conducted. We focus on two aspects: (i) mechanisms for locality-aware event dissemination and location-based filtering, as well as (ii) adaptation strategies and their realizations within the communication system. The former is required to derive suitable abstractions for *transition-enabled mechanisms* within a publish/subscribe system. The latter aids in defining requirements and strategies for the *seamless execution* of transitions.

Research Goal 2: *Realization of seamless mechanism transitions*

To realize adaptive event brokering, we require a methodology for the encapsulation of mechanisms and the execution of transitions. Our focus lies on the seamless execution of transitions between individual mechanisms in a publish/subscribe system.

By executing transitions, the system can benefit from the mechanism that is most suitable for the current environmental conditions. Given the dynamics of the mobile social applications, the overhead of a transition execution and the impact on system performance during the execution needs to be carefully considered. Consequently, for a seamless transition, state information gathered by an active mechanism needs to be utilized during the transition to another mechanism. In our contributions, this problem is tackled from two perspectives: (i) mechanism-specific within the domain of publish/subscribe [49, 50, 53–55] and (ii) generalized to ensure extensibility with new mechanisms and applicability to other domains [22, 51, 57, 65].

Research Goal 3: *Evaluation and characterization of mechanism transitions*

The effects of transitions between different mechanisms within the envisioned adaptive publish/subscribe system are to be measured. To this end, we contribute models for relevant features of the scenario of location-based mobile social applications, including mobility and workload models [52, 53, 57, 64]. To characterize the transition itself, we focus on the behavior of the system right during as well as after the execution of a transition. The evaluation should furthermore prove the applicability of our proposed generic design concepts for transition-enabled systems [22].

With the thesis focusing on mobile applications, we explicitly concentrate on the communication between a broker and mobile clients and on direct communication among mobile clients. We strive to support existing approaches for distributed broker networks within the proposed system design to benefit from the rich body of existing research. Consequently, we do not propose novel distributed broker networks.

When mobile users need to contribute own resources, e. g., for direct communication, incentives and protection against malicious clients are relevant research topics. While we pinpoint to relevant related work, we do not conduct own work in this area. The same holds true for privacy concerns regarding location-based services.

In this work, transitions are triggered reactively based on the currently observed state of the system. Combining model-based predictions of the future system behavior with additional external data sources is expected to further increase the efficiency of transitions. Realizing proactive planning of transitions is not a part of this thesis, but clearly constitutes a promising direction for future research.

1.4 Structure of the Thesis

Extending this brief introduction, we provide additional background on location-based mobile social applications, communication systems, event brokering, and mechanism transitions in Chap. 2. We discuss and classify related work in Chap. 3, addressing mobility support and means for adaptivity in publish/subscribe, further including a brief survey of mechanism transitions in other application domains.

Based on our discussion of related work, we propose BYPASS.KOM in Chap. 4. BYPASS.KOM is a modular framework allowing us to design, realize, and evaluate mechanism transitions in publish/subscribe systems. We discuss individual mechanisms, the transitions realized in the framework, and their execution strategies.

Our findings are generalized in Chap. 5, where we introduce the SIMONSTRATOR.KOM platform as an environment for the design and evaluation of mechanism transitions. We introduce core concepts of the platform and detail our transition-specific contributions before presenting our prototype realization of BYPASS.KOM.

We conduct an extensive evaluation of BYPASS.KOM and its mechanisms for location-based filtering and locality-aware event dissemination in Chap. 6. There, we specifically focus on the seamless execution of transitions by means of state transfer and different coordination concepts, showing detailed performance and cost characteristic for mechanism transitions. Additionally, we assess the general applicability of our methodology and design concepts for transition-enabled systems.

The thesis is concluded in Chap. 7 with a brief summary of the core contributions. Finally, we provide an outlook on potential future work.

References

1. Baldoni R, Marchetti C, Virgillito A, Vitenberg R (2005) Content-based publish-subscribe over structured overlay networks. In: Proceedings IEEE international conference on distributed computing systems (ICDCS). IEEE, pp 437–446
2. Bellavista P, Corradi A, Fanelli M, Foschini L (2012) A survey of context data distribution for mobile ubiquitous systems. In: ACM computing surveys (CSUR) 44.4, p 24
3. Brimicombe A, Li Y (2006) Mobile space-time envelopes for location- based services. Trans GIS 10(1):5–23
4. Burcea I, Jacobsen H-A (2003) L-ToPSS-push-oriented locationbased services. In: International workshop on technologies for E-services. Springer, pp 131–142
5. Caporuscio M, Carzaniga A, LWolf A (2003) Design and evaluation of a support service for mobile, wireless publish/subscribe applications. In: IEEE transactions on software engineering 29.12, pp 1059–1071
6. Carzaniga A, Rosenblum DS, Wolf AL (2000) Achieving scalability and expressiveness in an internet-scale event notification service. In: Proceedings ACM symposium on principles of distributed computing. ACM, pp 219–227
7. Carzaniga A, Rosenblum DS, Wolf AL (2001) Design and evaluation of a wide-area event notification service. In: ACM transactions on computer systems (TOCS) 19.3, pp 332–383
8. Chen X, Chen Y, Rao F (2003) An efficient spatial publish/-subscribe system for intelligent location-based services. In: Proceedings international workshop on distributed event-based Systems (DEBS). ACM, pp 1–6
9. Cheung AKY, Jacobsen H-A (2010) Publisher placement algorithms in content-based publish/subscribe. In: Proceedings international conference on distributed computing systems (ICDCS). IEEE, pp 653–664
10. Cilia M, Fiege L, Haul C, Zeidler A, Buchmann AP (2003) Looking into the past: enhancing mobile publish/subscribe middleware. In: Proceedings international workshop on distributed eventbased systems (DEBS). ACM, pp 1–8
11. Cisco (2016) Cisco Visual Networking Index: Forecast and Methodology, 2015–2020. http://www.cisco.com/c/en/us/solutions/collateral/service-provider/visual-networkingindex-vni/complete-white-paper-c11-481360.html. Accessed 8 Mar 2017
12. Cisco (2016) The Zettabyte Era: Trends and Analysis. http://www.cisco.com/c/en/us/solutions/collateral/service-provider/visual-networking-index-vni/vnihyperconnectivity-wp.html. Accessed 8 Mar 2017

13. Cugola G, Margara A, Migliavacca M (2009) Contextaware publish-subscribe: model, implementation, and evaluation. In: Proceedings IEEE symposium on computers and communications (ISCC). IEEE, pp 875–881
14. Cugola G, Picco GP (2006) REDS: a reconfigurable dispatching system. In: Proceedings international workshop on software engineering and middleware (SEM). ACM, pp 9–16
15. Dhar S, Varshney U (2011) Challenges and business models for mobile location-based services and advertising. Commun ACM 54(5):121–128
16. Dutz T (2017) Pervasive behavior interventions-using mobile devices for overcoming barriers for physical activity. PhD thesis. Technische Universität Darmstadt
17. Ericsson (2016) Ericsson Mobility Report: on the Pulse of the Networked Society. https://www.ericsson.com/res/docs/2016/ericsson-mobility-report-2016.pdf. Accessed 8 Mar 2017
18. Eugster PTh, Garbinato B, Holzer A (2005) Location-based publish/subscribe. In: Proceedings IEEE International symposium on network computing and applications. IEEE, pp 279–282
19. Eugster PTh, Felber PA, Guerraoui R, Kermarrec A-M (2003) The many faces of publish/subscribe. ACM Comput Surv (CSUR) 35.2, pp 114–131
20. Fiege L, Gärtner FC, Kasten O, Zeidler A (2003) Supporting mobility in content-based publish/subscribe middleware. In: Proceedings ACM/IFIP/USENIX 2003 international conference on middleware. Springer-Verlag New York, Inc., pp 103–122
21. Friedman R, Shulman AK (2013) A density-driven publish subscribe service for mobile ad-hoc networks. Ad Hoc Netw 11(1):522–540
22. Frömmgen A, Richerzhagen B, Rückert J, Hausheer D, Steinmetz R, Buchmann A (2015) Towards the description and execution of transitions in networked systems. In: Proceedings 9th international conference on autonomous infrastructure, management and security (AIMS). IFIP, pp 17–29
23. Frömmgen A, Hassan M, Kluge R, Mousavi M, Mühlhäuser M, Müller S, Schnee M, Stein M, Weckesser M (2016) Mechanism transitions: a new paradigm for a highly adaptive internet. Technical report Darmstadt. http://tuprints.ulb.tu-darmstadt.de/5370/
24. Gaddah A, Kunz T (2010) Extending mobility to publish/subscribe systems using a pro-active caching approach. Mobile Inf Syst 6(4):293–324
25. Girdzijauskas S, Chockler G, Vigfusson Y, Tock Y, Melamed R (2010) Magnet: practical subscription clustering for internet-scale publish/subscribe. In: Proceedings ACM international conference on distributed event-based systems (DEBS). ACM, pp 172–183
26. Gokhale A, Schmidt DC, Hoffert J (2011) Timely autonomic adaptation of publish/subscribe middleware in dynamic environments. Int J Adapt Resilient Auton Syst 2.4:1–24. https://doi.org/10.4018/jaras.2011100101
27. Google Inc. (2016) Google Project Fi - Network Coverage. https://fi.google.com/about/coverage/#network-of-networks
28. Holzer A, Eugster P, Garbinato B (2013) Evaluating implementation strategies for location-based multicast addressing. IEEE Trans Mobile Comput 12(5):855–867
29. Holzer A, Eugster P, Garbinato B (2012) Alps-adaptive location-based publish/subscribe. In: Computer networks 56.12, pp 2949–2962
30. Hu S, Muthusamy V, Li G, Jacobsen H-A (2009) Transactional mobility in distributed content-based publish/subscribe systems. In: Proceedings IEEE international conference on distributed computing systems (ICDCS). IEEE, pp 101–110
31. Huang Y, Garcia-Molina H (2001) Publish/subscribe in a mobile enviroment. In: Proceedings ACM international workshop on data engineering for wireless and mobile access. ACM, pp 27–34
32. Huang Y, Garcia-Molina H (2003) Publish/subscribe tree construction in wireless ad-hoc networks. In: Proceedings international conference on mobile data management. Springer, pp 122–140
33. Jayaram KR, Eugster P, Jayalath C (2013) Parametric contentbased publish/subscribe. In: ACM transactions on computer systems (TOCS) 31.2, p 4
34. Jayaram KR, Wang W, Eugster P (2015) Subscription normalization for effective content-based messaging. In: IEEE Transactions on parallel and distributed systems (TPDS) 26.11, pp 3184–3193

35. Lehn M, Rehner R, Buchmann A (2013) Distributed optimization of event dissemination exploiting interest clustering. In: Proceedings IEEE conference on local computer networks (LCN). IEEE, pp 328–331
36. Li G, Hou S, Jacobsen H-A (2005) A unified approach to routing, covering and merging in publish/subscribe systems based on modified binary decision diagrams. In: Proceedings IEEE international conference on distributed computing systems (ICDCS). IEEE, pp 447–457
37. Li G, Muthusamy V, Jacobsen H-A (2008) Adaptive contentbased routing in general overlay topologies. In: Proceedings ACM/IFIP/USENIX international conference on middleware. Springer-Verlag New York, Inc., pp 1–21
38. Liang S, Gao Y et al (2010) Real-time notification and improved situational awareness in fire emergencies using geospatial-based publish/subscribe. Int J Appl Earth Observ Geoinf 12(6):431–438
39. Moon S-C, Ko Y, Lee D (2007) A fast path recovery scheme for publish/subscribe in mobile ad hoc networks. In: Proceedings IEEE conference on computer and information technology (CIT). IEEE, pp 435–440
40. Mottola L, Cugola G, Picco GP (2008) A self-repairing tree topology enabling content-based routing in mobile ad hoc networks. IEEE Trans Mobile Comput 7(8):946–960
41. Mühl G, Fiege L, Buchmann A (2002) Filter similarities in content-based publish/subscribe systems. In: Trends in network and pervasive computing—ARCS 2002. Springer, pp 224–238
42. Muhl G, Ulbrich A, Herrman K (2004) Disseminating information to mobile clients using publish-subscribe. IEEE Int Comput 8(3):46–53
43. Muhl G, Fiege L, Gartner FC, Buchmann A (2002) Evaluating advanced routing algorithms for content-based publish/subscribe systems. In: Proceedings IEEE international symposium on modeling, analysis and simulation of computer and telecommunications systems (MASCOTS). IEEE, pp 167–176
44. Ottenwälder B, Koldehofe B, Rothermel K, Ramachandran U (2013) Migcep: operator migration for mobility driven distributed complex event processing. In: Proceedings ACM international conference on distributed event-based systems (DEBS). ACM, pp 183–194
45. Patel JA, Rivière É, Gupta I, Kermarrec A-M (2009) Rappel: Exploiting interest and network locality to improve fairness in publish-subscribe systems. Comput Netw 53(13):2304–2320
46. Pietzuch PR, Bacon JM (2002) Hermes: a distributed event-based middleware architecture. In: Proceedings conference on distributed computing systems workshops. IEEE, pp 611–618
47. Rebecchi F, Amorim MDD, Conan V, Passarella A, Bruno R, Conti M (2015) Data offloading techniques in cellular networks: a survey. IEEE Commun Surv Tutorials 17(2):580–603
48. Richerzhagen B, Koldehofe B, Steinmetz R (2015) Immense dynamism. German Res 37(2):24–27
49. Richerzhagen B, Steinmetz R (2013) Towards an adaptive publish/- subscribe approach supporting transitions. In: Proceedings 7th international conference on autonomous infrastructure, management and security (AIMS), PhD Workshop. IFIP, pp 84–87
50. Richerzhagen B, Stingl D, Hans R, Groß C, Steinmetz R (2014) Bypassing the cloud: peer-assisted event dissemination for augmented reality games. In: Proceedings 14th IEEE conference on peer-to- peer computing (P2P), pp 1–10
51. Richerzhagen B, Stingl D, Rückert J, Steinmetz R (2015) Simonstrator: simulation and prototyping platform for distributed mobile applications. In: Proceedings 8th international conference on simulation tools and techniques (SIMUTOOLS). ACM, pp 99–108
52. Richerzhagen B, Schiller M, Lehn M, Lapiner D, Steinmetz R (2015) Transition-enabled event dissemination for pervasive mobile multiplayer games. In: Proceedings 16th international symposium on a world of wireless, mobile and multimedia networks (WoWMoM). IEEE
53. Richerzhagen B, Wagener A, Richerzhagen N, Hark R, Steinmetz R (2016) A framework for publish/subscribe protocol transitions in mobile crowds. In: Proceedings 10th international conference on autonomous infrastructure, management and security (AIMS). IFIP, pp 1–14
54. Richerzhagen B, Richerzhagen N, Zobel J, Schönherr S, Koldehofe B, Steinmetz R (2016) Seamless transitions between filter schemes for location-based mobile applications. In: Proceedings 41st IEEE conference on local computer networks (LCN), pp 1–9

55. Richerzhagen B, Richerzhagen N, Schönherr S, Hark R, Steinmetz R (2016) Stateless gateways - reducing cellular traffic for event distribution in mobile social applications. In: Proceedings 25th international conference on computer communication and networks (ICCCN). IEEE, pp 1–9
56. Richerzhagen N, Stingl D, Richerzhagen B, Mauthe A, Steinmetz R (2015) Adaptive monitoring for mobile networks in challenging environments. In: Proceedings 24th international conference on computer communication and networks (ICCCN). IEEE, pp 1–8
57. Richerzhagen N, Richerzhagen B, Stingl D, Steinmetz R (2017) The human factor: a simulation environment for networked mobile social applications. In: Proceedings international conference on networked systems (NetSys). IEEE, pp 1–8
58. Rückert J, Richerzhagen B, Lidanski E, Steinmetz R, Hausheer D (2015) TopT: supporting flash crowd events in hybrid overlay-based live streaming. In: Proceedings 14th IFIP networking conference (Networking). IEEE, pp 1–9
59. Scellato S, Mascolo C, Musolesi M, Latora V (2010) Distance matters: geo-social metrics for online social networks. In: WOSN
60. Schiller J, Voisard A (2004) Location-based services. Elsevier
61. Segall B, Arnold D (1997) Elvin has left the building: a publish/subscribe notification service with quenching. In: Proceedings of AUUG. Brisbane, Australia, pp 243–255
62. Sivaharan T, Blair G, Coulson G (2005) Green: a configurable and re-configurable publish-subscribe middleware for perva sive computing. In: Proceedings international conferences on the move to meaningful internet systems (OTM). Springer, pp 732–749
63. Triantafillou P, Economides A (2004) Subscription summarization: a new paradigm for efficient publish/subscribe systems. In: Proceedings international conference on distributed computing systems (ICDCS). IEEE, pp 562–571
64. Long V, Nguyen P, Nahrstedt K, Richerzhagen B (2015) Characterizing and modeling people movement from mobile phone sensing traces. Pervasive Mobile Comput 17:220–235
65. Wichtlhuber M, Richerzhagen B, Rückert J, Hausheer D (2014) TRANSIT: supporting Transitions in peer-to-peer live video streaming. In: Proceedings IFIP networking conference (IFIP Networking). IEEE, pp 1–9
66. Wikipedia (2016) Transition (computer science)—Wikipedia, The Free Encyclopedia. https://en.wikipedia.org/wiki/Transition_(computer_science)
67. Xu Q, Erman J, Gerber A, Mao Z, Pang J, Venkataraman S (2011) Identifying diverse usage behaviors of smartphone apps. In: Proceedings ACM SIGCOMM conference on internet measurement (IM). ACM, pp 329–344
68. Yu M, Li G, Wang T, Feng J, Gong Z (2015) Efficient filtering algorithms for location-aware publish/subscribe. IEEE Trans Knowl Data Eng 27(4):950–963
69. Zeidler A (2005) A distributed publish/subscribe notification service for pervasive environments. PhD thesis. TU Darmstadt
70. Zhang D, Guo L, Nie L, Shao J, Wu S, Shen HT (2017) Targeted advertising in public transportation systems with quantitative evaluation. In: ACM transactions on information systems (TOIS) 35.3, p 20

Chapter 2
Background

In the following, we provide relevant background information on the scenario of location-based mobile social applications as motivated in Chap. 1. We start by discussing the respective applications, involved entities, and their interaction patterns in Sect. 2.1. The interaction patterns and entities of a mobile social application are modeled with the publish/subscribe communication paradigm, which is introduced in Sect. 2.2. This discussion is followed by a brief primer on relevant technical aspects of the underlying communication networks in Sect. 2.3. Lastly, we detail the concept of mechanism transitions in the context of communication systems, providing the foundation for the discussion of the state of the art in Chap. 3.

2.1 Location-Based Mobile Social Applications

As introduced in Chap. 1, location-based mobile social applications combine the characteristics of *location-based services* and *mobile applications*, with a focus on direct *social interaction* between users [33]. Before going into detail on the aspect of social interaction, we briefly summarize characteristics of mobile applications in general.

A mobile application is designed for a specific mobile Operating System (OS), usually through a dedicated *app store*. Examples for such stores include the Google Play Store or Apple's App Store for the Android and iOS Operating System, respectively. Being built against the programming interfaces and libraries of the respective OS, applications can benefit from a range of platform-specific features. Most notably, mobile applications can request access to the mobile device's sensors and contextual information. All major platforms provide specific programming interfaces allowing applications to request the current location of a device, information about its connectivity, and even higher-level information about a user's current activities.

B. Richerzhagen, *Mechanism Transitions in Publish/Subscribe Systems*,
Springer Theses, https://doi.org/10.1007/978-3-319-92570-7_2

A large fraction of the more traditional location-based applications simply provides user-friendly access to a geographical database or directory service. Examples include the search for nearby amenities like restaurants, ATMs, or bus stops [20]. The raw location data is further augmented with application-specific information, such as reviews of restaurants or the next departures at a nearby bus stop. Most notably, mobile users do not interact with each other via the application, but instead consume information from the central, cloud-based service back end. The back end usually utilizes a geographical database [2, 16] to store and retrieve content that is relevant at, or around, a specific location.

Information required at the service back end can either be generated by other users of the application (as with crowd-sourced restaurant reviews) or it can be fetched from external data sources and services (e.g., departure times). The number and type of utilized data sources heavily depends on the purpose and scope of the application. Consequently, more general purpose mapping applications such as Google Maps or Tripadvisor utilize and aggregate a large set of data sources.

In addition to the user's current location, context information such as upcoming calendar entries or current weather data is utilized to filter what is presented to the user even further. As most users keep their smartphone within arm's reach throughout the day, location-based applications have a great potential for personalization, especially if they fuse large amounts of data and context sources [39]. This becomes apparent from the recent trend towards personal assistants such as Apple's Siri or Google Now. Rather than solely reacting to users' input by fetching data from the cloud-based back end, these applications deploy a rather push-based data model. Here, information is gathered by the back end and actively presented to the user, if a set of contextual conditions is met. A simple example are location-based reminders, as provided by all the aforementioned applications: Here, a user can set a reminder not only at a given date and time, but at a specific location (e.g., a supermarket). As soon as the user approaches said location, the application notifies the user.

In addition to geographical databases and a centralized back end, distributed approaches for location-based search and retrieval of information have been proposed [11, 13, 19], based on the concept of Peer-to-Peer (P2P) search overlays [10]. Here, information is stored in a distributed fashion involving either user PCs in a fixed network [13, 14] or mobile devices in an ad hoc network [12].

Recently, location-based applications become increasingly focused on social interaction among users. Examples mentioned in the introduction are communities for sports and hobbies (e.g., *Sporty*, *PlayWith*, or *bvddy*) and location-based messaging applications (e.g., *Lokin*, *happn*), just to name a few. Here, the interaction among users exhibits certain locality characteristics, as verified in several user studies [9, 32, 38]. Users within close proximity tend to communicate with each other more often, depending on the target audience of an application. *Lokin*, for example, is an application provided by the Deutsche Bahn that enables communication with other users when commuting. It offers communication with other users on the same train ride, thereby exhibiting strong locality properties.

We refer to this effect as *locality of content and interest*. The importance of addressing interest locality has also been studied in the context of learning communities [28]. Here, Rensing et al. argue that the context and, most importantly, the location of users needs to be taken into account for both learners and content authors. The same effect is studied in [41, 42] for video streaming behavior in a campus network. Through an extensive measurement study, the authors confirm our intuitive understanding that video clips showing locally relevant content (e.g., recorded within proximity) exhibit higher local popularity among users. Locality properties as observed in these applications become even more significant for mobile augmented reality games [21], as discussed in the following section.

2.1.1 Mobile Augmented Reality Games

A prominent example for location-based mobile social applications are mobile augmented reality games. They exhibit characteristics of location-based applications, while at the same time focusing on live interaction between nearby users. In the following, we briefly discuss the core characteristics of such applications, as they have a significant impact on the utilization of the underlying communication system.

As the name suggests, an augmented reality game embeds its game-specific content into our physical world, rather than presenting a fully virtual environment as done in virtual reality applications and games. In a virtual reality application, a head-mounted appliance replaces a user's visual perception of the physical surroundings with the content generated by the respective application. The virtual reality application reacts to movements of the user's head and to additional sensor input, often collected by hand-held controllers. While this setup achieves a very immersive user experience through the manipulation of visual perception and additional use of audio, it also confines the actual physical mobility.

In contrast, augmented reality applications utilize a user's physical surroundings as canvas for the content displayed directly on the smartphone [40]. While some examples such as Google's Ingress simply utilize a map to display in-game content to users, others, such as Pokémon Go, make use of the smartphone camera, as illustrated in Fig. 2.1. In the latter case, content is directly displayed within a live video of the real-world surroundings. In contrast to virtual reality applications, a user actively explores content by moving around in the real world.

As the application displays content based on actual physical surroundings, user mobility is essential when interacting with the game. This characteristic motivated distinct mobile augmented reality applications for health [6]. Such applications aim to trigger the user to perform a given activity by including certain elements of games, referred to as a *gamification* of the respective application. A study by Althoff et al. shows that the mobile augmented reality game Pokémon Go leads to a moderate increase in daily activity [1]. However, according to another study by Howe et al., this effect attenuates after a few weeks. Still, the combination of augmented reality and physical mobility leads to interesting interaction characteristics among the users of a mobile augmented reality game, as discussed in the following section.

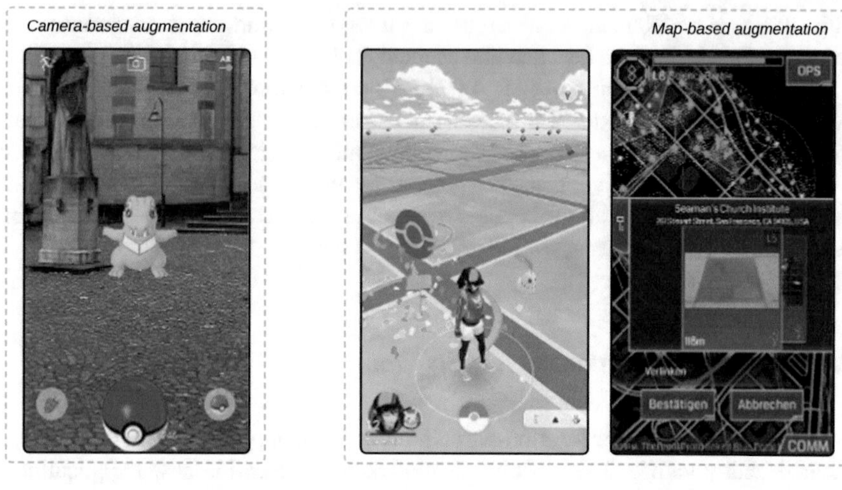

Fig. 2.1 Examples for camera-based and map-based augmentation (Images © Google)

2.1.2 Interaction and Attraction Characteristics

Users of mobile augmented reality games interact with (i) other users (in proximity and globally), and (ii) specific points of interest defined in the game. While the actual gameplay elements may vary depending on the game at hand, users interacting with the game access and manipulate a shared game world and its state. Consequently, actions of a single user influence the gameplay of other users. Figure 2.2 illustrates this interaction pattern between users and with Points of Interest (PoIs), as discussed in the following.

Users can interact with each other over larger distances through in-game chats. If users are within close proximity, game-specific direct interactions between these

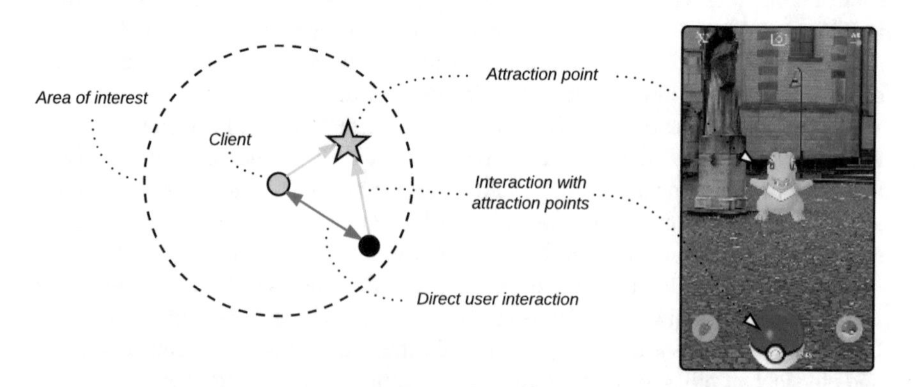

Fig. 2.2 Basic terms and interaction patterns in mobile augmented reality games

users become possible. The set of possible interactions can range from in-game transactions of goods or valuable items to interactive fights that rely heavily on the augmentation through the smartphone's camera. It is important to note that interaction between users does not necessarily have to involve the application itself. Instead, users might simply communicate verbally when in close proximity or by utilizing a third-party messaging application. Even if the communication itself does not take place within the application in such cases, it can still influence the game. These effects range from sudden movement of a local group of users to coordinated global actions with great impact on the game state. Examples include global operations in Google Ingress, where players around the world coordinate their actions to score points for their faction. In "Operation Green Marble", for example, thousands of players coordinated their actions to *win* the northern hemisphere for their faction in the augmented reality game Ingress.[1]

As mentioned, the real world is used as a canvas for the contents of the mobile augmented reality game. Consequently, real-world features such as streets, sights, and public places are incorporated into the gameplay as elements users can interact with. In Twostone [6], for example, public walkways are used to determine the structure of the playing field. In Google Ingress, public sights and historical buildings serve as so-called *portals*, in-game entities that users interact with. Within Pokémon Go, the probability of encountering and catching rare Pokémon is higher at specific locations, often at public parks or recreational areas. Independent of the game and regardless of the possible interactions, these elements can be classified as as PoIs. PoIs attract users due to the expected gain that is achieved when interacting with them in the game. Consequently, PoIs and their attractiveness within a game influence the mobility of users, and, in turn, on the utilization of the underlying communication system [17]. In this thesis, we use the term PoI and *attraction point* interchangeably.

To model the interaction between users and the interest-based communication in mobile social applications, we introduce the publish/subscribe communication paradigm in the following section.

2.2 Event Brokering and the Publish/Subscribe Paradigm

Figure 2.3 illustrates the general communication pattern of mobile social applications, modeled with the publish/subscribe paradigm. The cloud acts as a central *broker* for incoming information about users' actions, hereafter referred to as *events*. It is up to the broker to forward incoming events to all users that need to be notified of the particular event, a task commonly referred to as *event brokering*. Whether a user needs to be notified of a particular event can depend on many (usually application-specific) factors. For the case of a mobile augmented reality game, these factors include the user's physical distance to the originator of the event, for example.

[1] https://heise.de/-2224133 [Accessed March 8th, 2017].

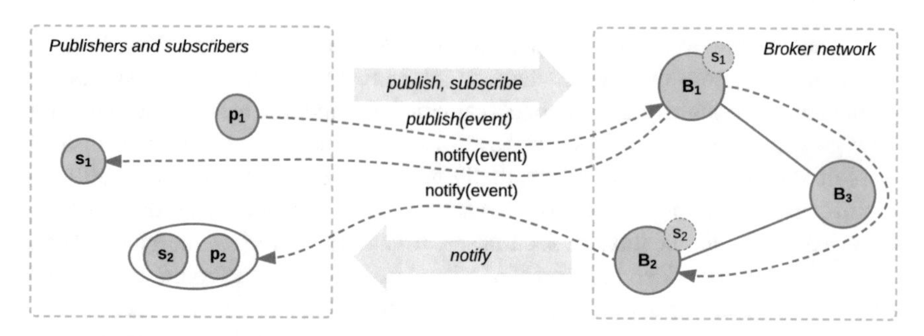

Fig. 2.3 Illustration of the publish/subscribe paradigm

The publish/subscribe paradigm offers a generic and commonly used protocol design for the realization of event brokering. It defines three distinct roles: *brokers*, *publishers*, and *subscribers* [7, 36]. These roles correspond to our entities, with the cloud, as already discussed, taking over the role of the broker. Users become publishers by generating events which they send to the broker. At the same time, users can also act as subscribers, as they are interested in a specific subset of events. The publish/subscribe paradigm defines how users state this interest: they subscribe at the broker, allowing the broker to filter incoming events against these subscriptions to determine the set of users that are to be notified.

By realizing event brokering with the publish/subscribe paradigm, senders and receivers of a specific event are decoupled with respect to three properties [7]. First, they are decoupled in space, meaning that the set of subscribers is not known by the publisher of an event. Instead, the publisher simply sends the event to the broker. Second, depending on the actual implementation, publishers and subscribers do not need to be online at the same time, decoupling them in time. Instead, the broker can be configured to store events and deliver these older events once a subscriber joins the system. Third, as publishers and subscribers communicate only with the broker and not with each other, they are decoupled with respect to synchronization.

The paradigm itself does not define the actual structure of subscriptions and the resulting filter operations at a broker. However, generic categories have emerged over the last years, differing in their expressiveness, intended application domain, and complexity. Existing publish/subscribe systems can be classified as (i) channel-based, (ii) attribute-based, or (iii) content-based. The respective scheme defines the basic structure of events and subscriptions, and thereby the expressiveness of the subscriptions that can be issued by an application.

The channel-based scheme supports subscriptions to pre-defined channels. These channels can be specified through strings or numerical identifiers, such as "`stock`". Some realizations of the channel-based scheme support tree-like nesting of channels by using an URL-like structure. This would allow subscriptions to "`stock`" that also cover events published to "`stock/google`" or "`stock/amazon`". Sometimes, channel-based schemes are also referred to as *topic-based*.

Attribute-based schemes enable subscriptions that filter based on a set of typed key-value pairs, referred to as attributes. Depending on the type of an attribute, different filter operations can be supported by the publish/subscribe system. For example, most attribute-based schemes support numerical filter operations like less-than, greater-than, and equal-to within a subscription. A subscription could filter, e.g., for events with a stock value greater than a given threshold: (stockvalue, 150, “>”). The syntax used for the example is explained in more detail in Sect. 4.2.1.

Content-based schemes aim to provide the highest level of expressiveness by allowing filters to operate directly on the application payload. Available filters commonly include text search or support for generic regular expressions, if the application payload is rather unstructured. Other approaches assume a specific structure of the payload and thereby increase the expressiveness of filters, while at the same time increasing the coupling between application and publish/subscribe system. Consequently, content-based publish/subscribe systems are often very much application-tailored.

Recently, especially for the scenario of mobile applications, a number of location-based schemes have been proposed. This includes generic context-based schemes. They enable filtering based on additional, time-dependent attributes such as the subscribers’ current locations. We discuss the respective filter schemes in detail in Sect. 3.1.3, given that their functionality constitutes an essential building block in a communication system for location-based mobile social applications.

However, a core characteristic of location-based subscriptions is their support for *interest mobility*, sometimes referred to as *logical mobility*. In contrast to static subscriptions resulting from the aforementioned schemes, a location-based subscription adapts itself by updating its filters to reflect the mobility of the subscriber. Consequently, the physical mobility of the subscriber is mapped to the publish/subscribe domain as a feature of the subscription scheme.

In the following, we provide a brief primer on the technical aspects of communication networks and, more specifically, mobile ad hoc networks. This section provides the technical foundation for the discussion of existing publish/subscribe systems and their suitability for mobile social applications in Chap. 3.

2.3 Communication Networks

Communication networks constitute the technical foundation for the aforementioned publish/subscribe systems. A communication network connects multiple end systems via a shared communication medium [26, 35]. In this thesis, we focus on the Internet as a prominent example of a communication network, often also referred to as a *computer network*. Computer networks are standardized by the Organization for Standardization (ISO) in the Open Systems Interconnection (OSI) model [26]. According to the OSI model, a computer network is abstracted into seven functional *layers*, as illustrated in Fig. 2.4. Each layer provides *services* to the upper layer

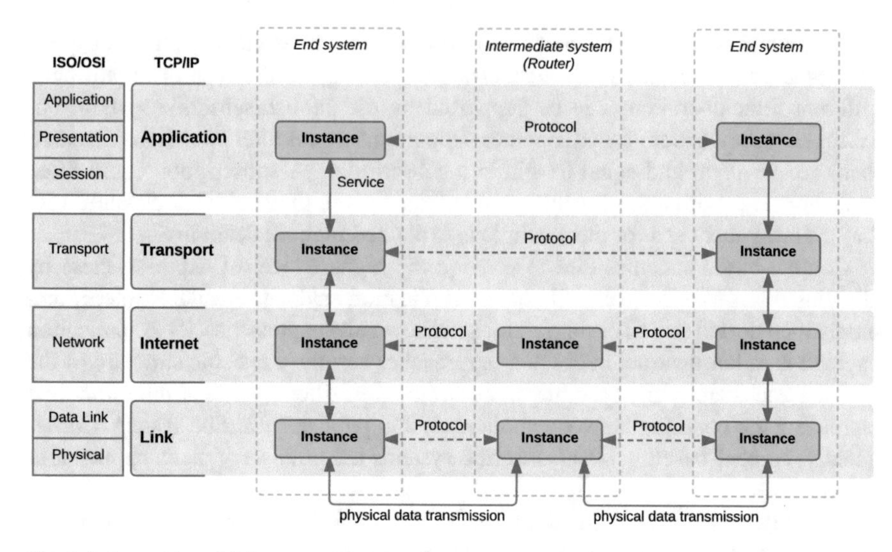

Fig. 2.4 Layered model for computer networks

and utilizes the respective services of the lower layer.[2] As illustrated in Fig. 2.4, instances of mechanisms realized on a given layer communicate with each other via well-defined *protocols*. Therefore, they exchange so-called Protocol Data Units (PDUs), with PDUs of a higher layer being concatenated by the layer below with a protocol header, a footer, or both. The actual transmission of data happens at the physical layer. The resulting flow of data is also shown in Fig. 2.4.

Regarding the Internet, a simplified version of the OSI model is usually considered: the TCP/IP model as shown in Fig. 2.4, named after the network protocol used in the Internet, IP, and the Transmission Control Protocol (TCP), a prominent transport protocol [26]. It is important to note that the network layer and all layers above operate end-to-end protocols between two systems. This is referred to as the *end-to-end principle*, according to which application-specific functionality is limited to the end hosts, whereas intermediate components are only utilized to establish the respective connection. This principle does not hold true for the Internet, especially if mobile devices and the respective cellular networks are considered, given that a number of intermediate components, e.g., Network Address Translation (NAT) boxes, intercept the respective protocols. However, for the mechanisms discussed in this thesis, we assume end-to-end connectivity, given that mechanisms for, e.g., NAT traversal can be applied to circumvent the respective middleboxes.

In the following, we model the Internet (including the cellular network) as a *black box* that simply enables communication between mobile devices and a (cloud-based)

[2]Technically, the ISO/OSI model specifies that service utilization is limited to adjacent layers. However, this rule is often (intentionally) violated to optimize for performance at the cost of loss of generality.

broker network with given Quality of Service (QoS) attributes. Considered QoS attributes include the latency, bandwidth, and reliability of the respective connection.

2.3.1 Communication Patterns

Based on this black box approach, we now discuss the communication pattern that results from applying the publish/subscribe paradigm to location-based mobile social applications as previously discussed. Mobile clients act as producers and consumers of data (i.e., publishers and subscribers of events). A mobile client sends an event to its assigned broker, thereby communicating in a one-to-one pattern. However, as multiple clients can be subscribed to the respective event, its distribution to the set of interested clients follows a one-to-many pattern.

The one-to-many pattern can be realized as a sequence of individual one-to-one transmissions (unicasts) from the broker to each subscriber. However, this leads to redundant data transmission within the communication network, consuming significant upload capacity of the respective broker. In a distributed broker network, multiple brokers might be involved in notifying the respective subscribers, leading to many-to-many communication. However, as later discussed in Sect. 3.2.1, clients are usually assigned to brokers based on their interest (i.e., their subscriptions) to increase routing efficiency within the broker network. Therefore, we assume that a single broker acts as source for the respective notifications.

To address this issue, multicast-based data dissemination can be utilized. Here, data intended for multiple recipients is transmitted only once by the broker, and the communication network itself takes care of delivering it to all interested clients. IP multicast has been proposed as a standardized protocol that realizes this behavior by addressing a group rather than individual clients. Clients can join the respective group to receive all data that is sent to the group's address. Group management is then performed by the intermediate routers via a dedicated protocol, the Internet Group Management Protocol (IGMP) in IPv4 and Multicast Listener Discovery (MLD) in IPv6. In practice, however, IP multicast is limited to a single Autonomous System (AS) and cannot be utilized for Internet-wide multicast.[3]

Given that we focus on location-based applications, we briefly discuss the geocast pattern. Instead of addressing individual clients or a group of clients, geocasting addresses a specific geographical coordinate or region. In an earlier work, we studied the utilization of geocasting protocols for mobile location-based services [12]. There, we relied on direct ad hoc communication between mobile devices. We introduce the resulting concept of a MANET in the following section.

[3]Even within an AS, it is usually not available for utilization by end users. This is due to the fact that accountability and security issues are not properly addressed by the group management protocols. However, IP multicast is commonly used for services that are offered directly by the respective provider, such as IP-based television (IPTV).

2.3.2 Mobile Ad Hoc Networks

The term Mobile Ad Hoc Network (MANET) is used to describe a wide range of self-organizing wireless networks with different envisioned application scenarios [4]. In general, a MANET is composed out of mobile entities that communicate with each other over a shared wireless medium in a self-organizing and decentralized fashion. Popular examples constitute Wireless Sensor Networks (WSNs) and, with increasing popularity, Vehicular Ad Hoc Networks (VANETs). According to Conti et al. [4], research interest and market relevance is especially high for vehicular applications and people-centric networks, with the latter being considered in this work from an architectural perspective. The authors further envision the combined and seamless utilization of multiple different MANET manifestations in future application scenarios, which we briefly address in Sect. 2.4.

Figure 2.5 illustrates key properties of a MANET, here formed by mobile devices (e.g., smartphones) as considered in this thesis. An individual device can communicate with other devices in range of the utilized wireless technology. For smartphone-based networks, usually Wi-Fi is utilized as underlying technology. More energy-efficient protocols, e.g., ZigBee or Bluetooth, are becoming increasingly popular for applications that do not require high throughput. To ease interoperability, MANETs often operate an IP network stack, with the exception of WSNs that are running on severely resource constrained devices. Different protocols exist to support end-to-end routing over a MANET [23, 30], however, their performance severely degrades with increasing client mobility and network size [3, 22].

To address this issue, a more decoupled approach to communication is usually used in MANETs. Multi-hop routing is addressed on the application layer relying only on single-hop unicasts or broadcasts on the lower layers. The resulting communication systems operate according to the multicast pattern, as discussed in more detail in Sect. 3.2.1. Depending on the application area, content is simply disseminated to all clients in the network. To this end, enabling efficient broadcasting in a MANET constitutes a research field on its own [31].

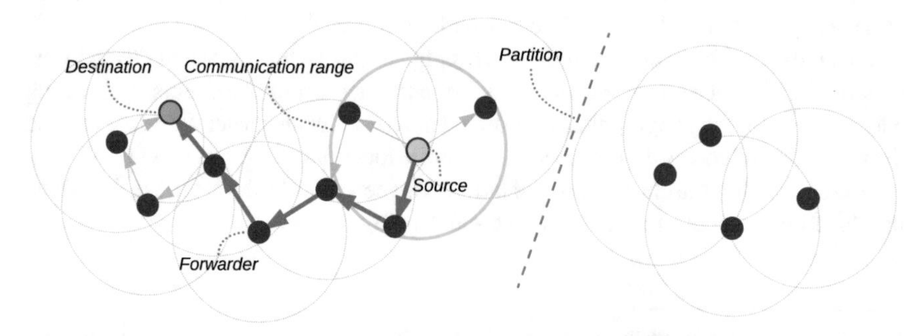

Fig. 2.5 Multi-hop communication in a Mobile Ad Hoc Network (MANET)

Figure 2.5 illustrates another issue arising in MANETs, especially if they are composed out of people carrying smartphones: partitions. Depending on the mobility and physical location of devices, areas of the network can become disconnected from other areas. These partitions might rejoin at a later point in time. This dynamic behavior of the underlying connections pose challenges to communication systems. At the same time, the behavior of participants in a MANET can also be utilized within the process of distributing content. In Delay-tolerant Networking (DTN), for example, content is intentionally stored at mobile devices and then forwarded at a later point in time, basically utilizing mobile devices as *data ferries*. While this approach utilizes client mobility to its advantage, the delay until content is finally delivered to an interested receiver increases substantially.

Several designs exist that propose to connect MANETs with the Internet [5]. The combination is especially interesting in the context of *mobile data offloading* [27]. Here, ad hoc connectivity among mobile devices is utilized to locally exchange data (e.g., chunks of a video or prefetched web content) instead of requesting it via the cellular network. As surveyed in [27], challenges arise in the identification of potential sources and their coordination with and without central infrastructure. Additionally, potentially selfish mobile clients need to be incentivized to contribute their resources by acting as a *gateway* or relay for other clients [24, 25].

Especially in MANETs (but also in fixed networks), adaptivity to dynamic environmental and load conditions is a key requirement. This requirement is addressed with the design concept of mechanism transitions for adaptive communication systems, as introduced in the following section.

2.4 Concept of Mechanism Transitions

The general concept of a transition between individual mechanisms within a communication system is formalized by members of the "MAKI"[4] in [8, 29, 34, 37]. In the following, we provide a summary of transition-enabled communication systems. A communication system consists of multiple mechanisms, operating on different layers of the networking stack, as previously discussed in Sect. 2.3. TCP, for example, is a mechanism on the transport layer. In today's Internet, a multitude of mechanisms exist for a specific function: in addition to TCP, for example, there are several other transport protocols such as UDP or the Stream Control Transmission Protocol (SCTP). In MAKI, a group of mechanisms that provide exchangeable functionality is defined as a *multi-mechanism* [8], as illustrated in Fig. 2.6.

Although each mechanism within a multi-mechanism realizes the same functionality (for example, packet transport), mechanisms exhibit different performance and cost characteristics. These characteristics depend on the application scenario, the available resources, and the constraints imposed by the current environmen-

[4] www.maki.tu-darmstadt.de [Accessed March 8th, 2017].

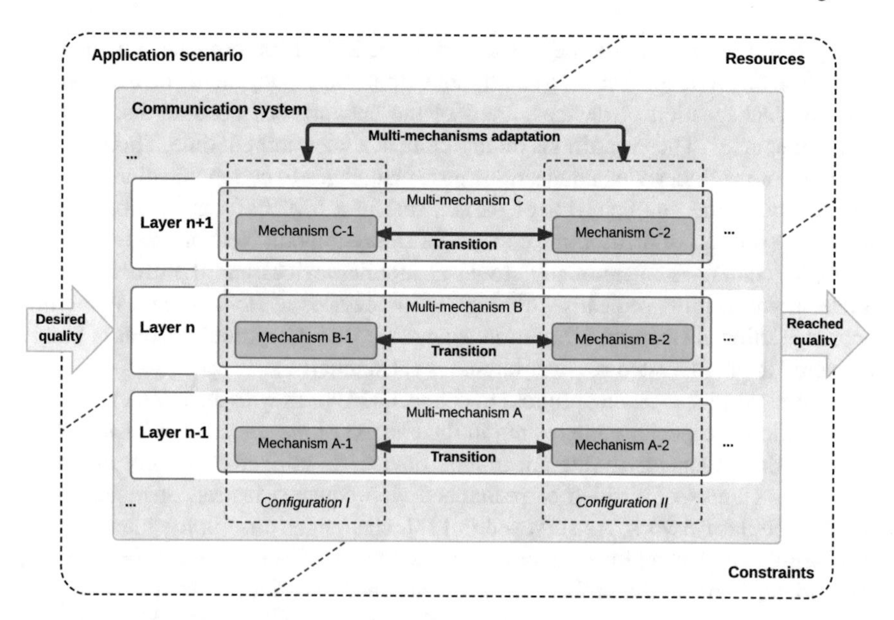

Fig. 2.6 Mechanisms and multi-mechanisms within a communication system

tal conditions. Consequently, to cope with dynamic environments and application requirements, mechanisms need to be adaptive. Historically, this has led to an ever increasing number of new mechanisms.

Instead of focusing on the design of new mechanisms, MAKI aims to utilize existing mechanisms in a highly adaptive fashion. By realizing transitions between mechanisms within a multi-mechanism, the currently running mechanism can be exchanged seamlessly during operation of the communication system. By coordinating such transitions across multiple functional layers and the respective multi-mechanisms, a *multi-mechanisms adaptation* is achieved. Such a coordinated execution of transitions affecting multiple mechanisms leads to the systems switching from one configuration to another.

The set of suitable configurations is determined by the application scenario, the environmental constraints, available resources, and the desired quality. Available resources do not only pose a limitation to the transition-enabled system. Instead, they can also aid in the execution of transitions. To this end, MAKI considers the utilization of SDN for the coordination and execution of transitions affecting the available resources [15]. Thereby, the transition-enabled communication system can actively alter and influence available resources.

Transitions can affect multiple layers and, given the distributed nature of communication systems, have different regional scopes within a system. For their coordination, a distributed control cycle is envisioned within MAKI. This process follows the

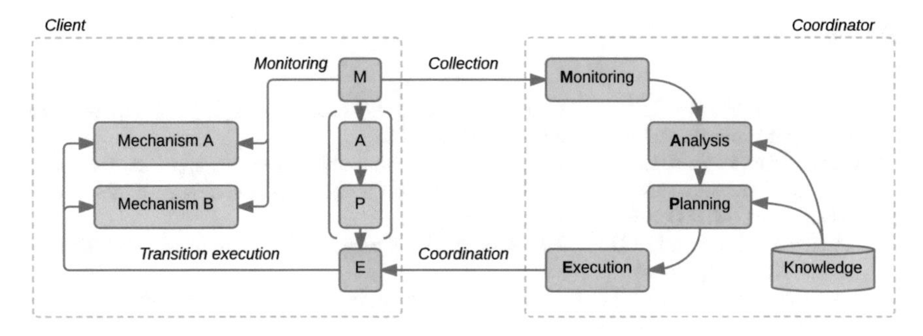

Fig. 2.7 The MAPE-cycle for the adaptation of (distributed) systems

steps of the well-known MAPE-cycle [18] of *monitoring* a system's or mechanism's state, *analyzing* the gathered information, *planning* suitable transitions, and finally *executing* the transitions in the network, as illustrated in Fig. 2.7. The example shows a central coordinator that performs the analysis and planning steps of the cycle. Analysis and planning might involve a *knowledge* component containing, e.g., a model of the system that is to be coordinated or information about previously observed conditions and the respective adaptations. Monitoring involves collecting relevant information from the clients that are to be controlled as well as environmental information collected from other sources (not illustrated). Once the coordinator comes to a conclusion about the adaptation that is to be performed (or the set of transitions in our case), it notifies the respective execution component at the client. This component then takes care of the execution of transitions to adapt the system accordingly by interacting with the respective mechanisms.

As indicated in the example, the MAPE-cycle can also be realized in a fully distributed manner, with each client performing its analysis and planning steps based on local or regional knowledge.

Research in the Collaborative Research Centre "MAKI" addresses distinct aspects of adaptation mechanisms, communication mechanisms, and a design methodology for transition-enabled communication systems.Within this work, we apply the concept of mechanism transitions to the specific domain of publish/subscribe systems, as discussed in Chap. 4. We further contribute to the generalized design methodology for transition-enabled communication systems by proposing domain-independent abstractions of the concepts proposed for publish/subscribe mechanisms in Chap. 5. We evaluate the impact of coordinated mechanism transitions on the performance and behavior of a communication system, specifically focusing on the time right during and after execution. In the following chapter, we survey state of the art mechanisms and adaptation principles that motivate our design and implementation of transitions in publish/subscribe systems.

References

1. Althoff T, White RW, Horvitz E (2016) Influence of Pokémon Go on physical activity: study and implications. In: arXiv preprint arXiv:1610.02085
2. Chen Y, Rao F, Yu F, Liu F (2003) CAMEL: a moving object database approach for intelligent location aware services. In: Proceedings IEEE international conference on mobile data management (MDM). Springer, pp 331–334
3. Chlamtac I, Conti M, Liu JJ-N (2003) Mobile ad hoc networking: imperatives and challenges. Ad hoc Netw 1(1):13–64
4. Conti M, Giordano S (2014) Mobile ad hoc networking: milestones, challenges, and new research directions. IEEE Commun. Mag. 52(1):85–96
5. Ding S (2008) A survey on integrating MANETs with the Internet: challenges and designs. Comput Commun 31(14):3537–3551
6. Dutz T (2017) Pervasive behavior interventions-using mobile devices for overcoming barriers for physical activity. PhD thesis. Technische Universität Darmstadt
7. Eugster PTh, Felber PA, Guerraoui R, Kermarrec A-M (2003) The many faces of publish/subscribe. In: ACM computing surveys (CSUR) 35.2, pp 114–131
8. Frömmgen A, Hassan M, Kluge R, Mousavi M, Mühlhäuser M, Müller S, Schnee M, Stein M, Weckesser M (2016) Mechanism transitions: a new paradigm for a highly adaptive internet. Technical report Darmstadt. http://tuprints.ulb.tu-darmstadt.de/5370/
9. Gao H, Tang J, Liu H (2012) Exploring social-historical ties on location-based social networks. In: ICWSM
10. Groß C, Richerzhagen B, Lehn M (2013) Structured search overlays. In: Benchmarking peer-to-peer systems. Springer, Berlin Heidelberg, pp 49–67
11. Groß C, Stingl D, Richerzhagen B, Hemel A, Steinmetz R, Hausheer D (2012) Geodemlia: a robust peer-to-peer overlay supporting location-based search. In: Proceedings 12th IEEE international conference on peer-to-peer computing (P2P). IEEE, pp 25–36
12. Groß C, Stingl D, Gottron C, Richerzhagen B, Münker C, Hausheer D (2012) Harnessing mobile ad hoc communication for decentralized location-based services. Technical report Peer-to- Peer Systems Engineering Lab, TU Darmstadt, Germany
13. Groß C, Richerzhagen B, Stingl D, Munker C, Hausheer D, Steinmetz R (2013) Geodemlia: persistent storage and reliable search for peer-to-peer location-based services. In: Proceedings 13th IEEE international conference on peer-to-peer computing (P2P). IEEE, pp 1–2
14. Groß C, Richerzhagen B, Stingl D, Weber J, Hausheer D, Steinmetz R (2013) Geoswarm: a multi-source download scheme for peer-to-peer location-based services. In: Proceedings 13th IEEE international conference on peer-to-peer computing. IEEE, pp 1–10
15. Heuschkel J, Stein M, Wang L, Mühlhäuser M (2017) Beyond the core: enabling software-defined control at the network edge. In: Proceedings international conference on networked systems (NetSys). IEEE, pp 1–6
16. Ilarri S, Mena E, Illarramendi A (2010) Location-dependent query processing: Where we are and where we are Heading. In: ACM computing surveys (CSUR) 42.3 (2010), p 12
17. Kayastha N, Niyato D, Wang P, Hossain E (2011) Applications, architectures, and protocol design issues for mobile social networks: a survey. Proc IEEE 99(12):2130–2158
18. Kephart JO, Chess DM (2003) The vision of autonomic computing. Computer 36(1):41–50
19. Kovacevic A (2009) Peer-to-peer location-based search: engineering a novel peer-to-peer overlay network. PhD thesis. TU Darmstadt
20. Küpper A (2005) Location-based Services: Fundamentals and Operation. Wiley
21. Lehn M (2016) InterestCast: adaptive event dissemination for interactive real-time applications. PhD thesis. Technische Universität Darmstadt
22. Li J, Blake C, SJ De Couto D, Lee HI, Morris R (2001) Capacity of ad hoc wireless networks. In: Proceedings international conference on mobile computing and networking. ACM, pp 61–69
23. Mauve M, Widmer J, Hartenstein H (2001) A survey on positionbased routing in mobile ad hoc networks. IEEE Netw 15(6):30–39

24. Mousavi M, Al-Shatri H, Hinz O, Klein A (2016) Incorporating user willingness for message forwarding in multi-hop content distribution scenarios. In: Proceedings international ITG workshop on smart antennas (WSA). VDE, pp 1–6
25. Mousavi M, Müller S, Al-Shatri H, Freisleben B, Klein A (2016) Multi-hop data dissemination with selfish nodes: Optimal decision and fair cost allocation based on the Shapley value. In: Proceedings IEEE international conference on communications (ICC). IEEE, pp 1–6
26. Peterson LL, Davie BS (2007) Computer networks: a systems approach. Elsevier
27. Rebecchi F, Amorim MDD, Conan V, Passarella A, Bruno R, Conti M (2015) Data offloading techniques in cellular networks: a survey. IEEE Commun Surv Tutorials 17(2):580–603
28. Rensing C, Tittel S, Steinmetz R (2012) Location-based services for technology enhanced learning and teaching. In: Software service and application engineering. Springer, pp 165–179
29. Richerzhagen B, Koldehofe B, Steinmetz R (2015) Immense dynamism. German Res 37(2):24–27
30. Royer EM, Toh C-K (1999) A review of current routing protocols for ad hoc mobile wireless networks. IEEE Pers Commun 6(2):46–55
31. Ruiz P, Bouvry P (2015) Survey on broadcast algorithms for mobile ad hoc networks. In: ACM computing surveys (CSUR) 48.1, p 8
32. Scellato S, Mascolo C, Musolesi M, Latora V (2010) Distance matters: geo-social metrics for online social networks. In: WOSN
33. Schiller J, Voisard A (2004) Location-based services. Elsevier
34. Steinmetz R, Holloway M, Koldehofe B, Richerzhagen B, Richerzhagen N (2015) Towards future internet communications—role of scalable adaptive mechanisms. In: ACADEMIA EUROPAEA, pp 59–61
35. Tanenbaum A (2002) Computer networks. 4th. Prentice Hall Professional Technical Reference. ISBN: 0130661023
36. Tarkoma S (2012) Publish/subscribe systems: design and principles. Wiley
37. Wikipedia (2016) Transition (computer science)—Wikipedia, The Free Encyclopedia. https://en.wikipedia.org/wiki/Transition_(computer_science)
38. Xu Q, Erman J, Gerber A, Mao Z, Pang J, Venkataraman S (2011) Identifying diverse usage behaviors of smartphone apps. In: Proceedings ACM SIGCOMM conference on internet measurement (IM). ACM, pp 329–344
39. Yin H, Sun Y, Cui B, Hu Z, Chen L (2013) LCARS: a location-content-aware recommender system. In: Proceedings ACM SIGKDD international conference on knowledge discovery and data mining. ACM, pp 221–229
40. Zaid F, Mogre PS, Reinhardt A, Costantini D, Steinmetz R (2010) iVu. KOM: A Framework for Viewer-centric Mobile Locationbased Services. In: PIK-Praxis der Informationsverarbeitung und Kommunikation 33.4, pp 284–290
41. Zink M, Suh K, Gu Y, Kurose J (2008) Watch global, cache local: YouTube network traffic at a campus network: measurements and implications. In: Electronic imaging 2008. International society for optics and photonics, pp 681805–681805
42. Zink M, Kyoungwon Suh YG, Kurose J (2009) Characteristics of YouTube network traffic at a campus network-measurements, models, and implications. Comput Netw 53(4):501–514

Chapter 3
State of the Art

In this chapter, we discuss the state of the art in adaptive event brokering for location-based mobile social applications. We discuss how existing mechanisms and systems address three distinct questions resulting from the scenario of mobile social applications: (i) *how do publish/subscribe mechanisms address client, broker, and interest mobility?* (Sect. 3.1), (ii) *how do they achieve adaptivity in dynamic and heterogeneous environments?* (Sect. 3.2), and (iii) *how are transitions utilized to adapt communication systems?* (Sect. 3.3). The individual findings are summarized in Sect. 3.4, highlighting the research gap addressed in our work.

3.1 Mobility Support in Publish/Subscribe Systems

Mobility is an inherent and core characteristic of mobile social applications. However, in location-based applications, mobility needs to be considered with respect to two different perspectives: physical and regarding the interest. The *physical mobility* of clients and devices needs to be considered in the design of the publish/subscribe system. While mechanisms such as a client handover can ease the handling of mobile clients in publish/subscribe systems, additional means might be required to also deal with intermittent connection failures or quality degradations of the cellular connection caused by mobility. We discuss mechanisms that deal with physical mobility of clients (i.e., publishers and subscribers) in Sect. 3.1.1, before also considering mobile brokers in the context of MANETs in Sect. 3.1.2. This provides the foundation for our contributions on the localityspsaware dissemination of events.

The client interest in content in location-based mobile social applications also depends on the physical location and, thus, is subject to mobility. The resulting *interest mobility* affects the inner workings of the publish/subscribe filtering mechanisms.

B. Richerzhagen, *Mechanism Transitions in Publish/Subscribe Systems*,
Springer Theses, https://doi.org/10.1007/978-3-319-92570-7_3

While already well known in the context of interest management for Networked Virtual Environments [58], adapting the respective concepts to a mobile scenario is a challenging issue. We discuss the state of the art in supporting interest mobility in Sect. 3.1.3, laying the foundation for our contributions on location-based filtering.

3.1.1 Producer and Consumer Mobility

A large fraction of publish/subscribe systems was initially designed for fixed networks and, consequently, non-mobile producers and consumers of events. Examples include research prototypes such as REBECA [67], HERMES [74], SIENA [10], PADRES [53], and commercially available middleware solutions like the CORBA event and notification services [87]. Besides, the JINI framework [2] with Apache River[1] as prominent reference implementation is to be mentioned. The main focus of these projects lies on the organization of brokers, the representation of subscriptions (often referred to as the *subscription language*), and the resulting expressiveness of the filtering process, all within a rather static environment assuming fixed networks.

However, due to the increasing number of handheld devices and new application domains, the need to support mobile producers and consumers arises. Decentralized broker networks aim at increasing the locality of event processing by bundling subscriptions [97] with similar content on a single broker, as discussed in more detail in Sect. 3.2. Consequently, associating clients to a broker based on their current location is an intuitive design goal, especially as in location-based services physical location and interest in content are intertwined. Given that clients are mobile, the broker network needs to support client handovers between different brokers. Such a handover process can be realized with various different mechanisms [9, 13, 14, 23, 26, 35, 66, 71], depending on the underlying system model and the application scenario, as explained in the following.

In [66], Mühl et al. extend the REBECA prototype [67] to support client mobility through *persistent connections* that handle intermittent connection failures or re-associations of mobile clients transparently to the client. To this end, the state associated to a connection (client subscriptions and the transmission queue for events) is exchanged between brokers whenever a mobile client connects to a new broker. At the same time, the routing paths within the broker network are adapted accordingly to ensure that relevant events are continued to be forwarded to the client.

Mobile clients might intentionally drop their connection to the publish/subscribe system to save bandwidth and/or battery power while the application is not actively used. Whenever the connection is re-established, the application needs to retrieve at least a subset of the previously published events for the client's current location to provide meaningful information. Cilia et al. propose an extension of REBECA's subscription model in [14] to support bootstrapping of mobile clients by replaying past events. The authors study two distinct approaches to this goal. The first approach

[1] http://river.apache.org [Accessed March 8th, 2017].

enables clients to state a timespan when actively subscribing. The timespan determines how long events need to be stored at a broker to be replayed to a mobile client when rejoining. This requires changes to the subscription model used in REBECA and sufficient caching capabilities of brokers. The second approach introduces a virtual client acting as a cache for events on behalf of the potentially disconnected mobile client. A similar concept is proposed by Caporuscio et al. in [9] and by Fiege et al. in [23], where the authors introduce proxies acting as caches whenever a mobile client is disconnected. Events stored within a proxy are sent to the client once the connection is re-established.

The aforementioned approaches are reactive and migrate clients in the event of connectivity failures or re-associations. A proactive approach for mobile clients is presented by Ottenwälder et al. in [71, 72] in the scope of a Complex Event Processing (CEP) system for automotive scenarios. Within the CEP system, the operators and their associated state information need to be migrated to brokers nearby the client's current location to maintain low delivery delays. The authors propose proactive planning of migrations based on expected client mobility and the cost of dependent migrations. Gaddah et al. propose a similar proactive scheme for mobility management in publish/subscribe systems in [26]. Here, a mobile client's subscriptions are forwarded to nearby brokers based on a neighborhood graph. At these brokers, the subscriptions remain in a passive state, meaning that they are not considered during the filtering process of the respective brokers. The handover operation activates the subscriptions at the broker that is subsequently contacted by the client as consequence of the movement. Cheung et al. further propose two algorithms to proactively migrate clients based on their current work in [13]. The handover process as such has been modeled as a transaction in [35], involving not only the edge brokers but also intermediate brokers in the broker network.

In addition to the support for client handover and migration, other mechanisms have been proposed to optimize communication with mobile consumers and producers. As a significant first step to reduce the communication overhead within an event brokering system, the brokers themselves are organized in a hierarchical and geographically distributed fashion. Thereby, processing capabilities are moved closer towards the producers and consumers. Saurez et al. propose the utilization of so-called *Foglets* within a highly distributed and localized broker platform in [85]. The authors envision a multi-tiered hierarchy, where edge and core routers possess computational capabilities to act as brokers for events produced and consumed by clients. In the proposed information model, each data object is associated with a location and filtered based on that information. The proposed system utilizes the migration model proposed by Ottenwälder et al. in [71] to deal with mobile clients.

Within the aforementioned extension of REBECA presented by Mühl et al. in [66], the authors also propose a mechanism that limits the rate of events sent to a mobile consumer. Therefore, events are categorized based on their expected importance for the client application, requiring application-specific knowledge. In case the event rate exceeds a configurable threshold, only the most relevant events are sent to the mobile client. This is a technique also applied to event brokering in wireless sensor networks [39]. The savings in terms of reduced cellular network utilization come at the cost of incomplete knowledge at the client.

In [36], Huang et al. utilize the concept of *quenching*, initially proposed for fixed broker networks by Segall et al. in [86], to reduce the number of event transmissions. The broker sends an aggregate of all currently active subscriptions to mobile producers. This enables producers to locally determine whether a new event needs to be sent to the broker network at all. However, the scheme can only operate efficiently if the set of subscriptions can be aggregated and subscriptions are static. In case of dynamic subscriptions, the aggregates would change frequently, leading to increased traffic overhead. At the same time, quenching requires mobile clients to (at least partially) execute a broker's task of filtering, leading to increased complexity of the client implementation. The concept of quenching is utilized in our work on context-based subscriptions, as later discussed in Sect. 3.1.3.

3.1.2 Broker Mobility in Mobile Ad Hoc Networks

The brokers themselves can also be subject to mobility if the publish/subscribe system is deployed on a Mobile Ad Hoc Network. In MANETs, the role of the broker is distributed among the mobile devices. To reduce the communication overhead caused by wide dissemination of events, publish/subscribe systems for MANETs aim to exploit locality properties. This is true even if end-to-end connectivity is provided by a routing mechanism, as sending messages via multiple hops is a costly operation and significantly reduces the overall throughput in dense networks [55]. Consequently, most publish/subscribe systems for MANETs do not assume any underlying routing algorithm. Instead, they rely on custom mechanisms to create routing structures that cope with the dynamics and expected size of the MANET for the target application.

Tree-based overlay approaches for publish/subscribe have been proposed in [25, 37, 63, 64]. Additionally, many general-purpose broadcast protocols for MANETs utilize tree-like routing structures. How the respective tree structure is created, differs slightly between approaches. Ruiz et al. provide an excellent survey on the matter in [84]. When utilized for publish/subscribe systems, the tree usually connects the clients that are currently acting as brokers. Similar to the construction of the tree, the selection process for brokers differs between approaches. Friedman et al. propose to select brokers based on the neighborhood density in [25]. The authors utilize a gradient-based routing scheme to forward events and subscriptions from non-broker clients to their nearest broker. Brokers communicate with each other using a *density-biased walk* that leads to an implicit spanning tree between all brokers. While this works reasonably well in rather static scenarios with a low event rate, increased client mobility requires frequent handovers of broker functionality between clients and aggressive tree management to ensure deliveries. To this end, Mitra et al. utilize a mobility prediction model to adapt the tree structure to the predicted movement of clients in [62]. Still, the increasing overhead in mobile scenarios combined with a decreasing performance especially under higher event rates as reported in [64] renders tree-based approaches unsuitable for the scenario of location-based mobile social applications. Additionally, locality in client interaction is usually not considered, as the systems proposed in [25, 37, 63, 64] forward an event to *all* brokers. Consequently, hierarchical approaches have been proposed to utilize different routing schemes for local and global communication. Yoo et al. , for example, combine regionally constrained flooding within clusters of clients with tree-based forwarding between those clusters in [104].

Meier et al. propose STEAM [60, 61], a locality-aware approach to event brokering in MANETs. STEAM utilizes a proximity-based approach to event forwarding and processing, thereby combining locality-aware dissemination with location-based subscriptions discussed in more detail in Sect. 3.1.3. By stating an explicit range for the validity of events, their dissemination in the network is limited. Similarly, Costa et al. propose to utilize an unstructured P2P overlay to disseminate events in a semi-probabilistic fashion [15]. Clients locally rely on a partial view of current subscriptions to decide whether to forward events or not. Other probabilistic schemes utilize Bloom filters [6] to summarize subscriptions, based on which forwarding decisions are made [103]. By using probabilistic data structures such as Bloom filters, the coordination overhead can be reduced significantly. This concept is also commonly applied in scenarios with resource constrained devices, such as sensor networks [90]. Datta et al. [19] argue that completeness is not necessarily a requirement in real-life civilian applications and propose a content-driven algorithm for data dissemination. However, the DTN-like nature of the algorithm renders it unsuitable for timely event propagation. A similar approach is also proposed by Costa et al. in [16], realizing publish/subscribe by forwarding messages based on predicted social interactions of the respective clients.

In the extreme case, each individual subscriber also acts as a broker for its own subscriptions, with events being disseminated to all clients in the network. This is also referred to as information diffusion and a plethora of protocols have been proposed, as surveyed by Ruiz et al. in [84]. We limit our discussion to representative examples that are later utilized as transition-enabled mechanisms within our work. In general, diffusion mechanisms cannot guarantee reliable delivery of events due to the possibility of network partitions in a MANET. Khelil et al. propose HYPERG [44], a hybrid gossiping-based dissemination approach that adapts to MANETs with different client densities and mobility characteristics by altering the forwarding probability accordingly. Each client periodically broadcasts a beacon that is then used by neighboring clients to estimate the current density. The authors specifically focus on handling network partitions by implementing a heuristic to retransmit messages if a partition is detected.

Holzer et al. add information about a client's current location to outgoing broadcasts in PLAN- B [33]. When receiving a message, clients calculate their distance to the respective sender. A hesitation time and forwarding probability is chosen based on the calculated distance, such that clients further away are preferred as forwarders. PLAN- B disseminates an event to all clients within a network. Besides, geospatial information can be utilized to forward events to a specific region rather than all clients in the network. Navas and Ko [48, 69] coined the term geocasting for the respective protocols. When utilized for location-based applications, dissemination schemes based on geospatial information can further help in lowering the load on the network [29], potentially supporting large-scale deployments as considered in [70]. An extensive survey on existing protocols is provided by Maihofer et al. in [59]. In their context-based publish/subscribe system for MANETs [24], Frey et al. rely on a geocasting algorithm to disseminate events based on the associated spatial context. Holzer et al. conduct an evaluation of different algorithms for location-based multicasting [31]. They propose a hybrid approach that disseminates queries (similar to subscriptions in our scenario) and events using a geographically scoped gossiping algorithm earlier proposed in [32]. Subsequently, matching occurs in a rendezvous-like fashion, similar to BUBBLESTORM proposed for fixed networks by Terpstra et al. [96]. Wang et al. [99] apply the concept of geospatial routing to vehicular networks [8]. The authors explicitly focus on the direction of movement rather than the exact location. This is motivated by the fact that information disseminated in an automotive scenario is mostly relevant on a specific lane or within a specific direction of travel.

A pure MANET-based approach is not suitable for the scenario considered in this work, as we still need to maintain global connectivity among users of a location-based mobile social application. However, the utilization of direct ad hoc communication and strategies for the distributed coordination of mobile devices are vital in achieving strong locality in event brokering. In [51], Leontiadis et al. propose a hybrid system for vehicular scenarios, where content is routed into a target area via infrastructure networks (if available). Once content reached the target area, it is continuously propagated in an ad hoc fashion for a specified amount of time. Hybrid approaches usually aim at offloading the cellular connection, as recently surveyed by Rebecchi et al. in [76]. Offloading via gateway-based infusion of content into a target area requires a selection procedure for gateways as outlined in [15, 57], for example. We utilize gateways for our hybrid dissemination mechanisms as later discussed in Chap. 4.

3.1.3 Interest Mobility Through Location-Based Filtering

In location-based applications, interest in a specific event also depends on the current location of the mobile subscriber and, potentially, on the location of the producer. Simply encoding this information within subscriptions can lead to significant overhead, as clients need to frequently alter their subscriptions to reflect their current location. According to Cugola et al. [17] and Fiege et al. [23], for example, it is inevitable to treat contextual information explicitly during event brokering. According to the authors, the flexibility of the system and its efficiency in a dynamic scenario is otherwise severely limited. In the following, we discuss how location-based filtering can be enabled (i) by extending existing attribute- or content-based publish/subscribe systems as proposed in [7, 17, 23, 26, 41, 106], and (ii) by introducing new subscription models for context-based filters [3, 5, 12, 22, 43, 56, 105].

Fiege et al. discuss the consequences of logical mobility for the content-based publish/subscribe system REBECA in [23, 106]. The authors propose a technique to limit the overhead caused in the broker network when propagating subscription updates as a consequence of mobility. Therefore, the most accurate filter is only maintained at the broker that is responsible for the respective client. Filters propagated to other brokers in the broker network are less restrictive and forward events destined to locations that might become relevant for the client in the near future. Similar to the approach for handling physical mobility presented in [26], the authors motivate using a neighborhood graph that contains potential future logical locations to determine suitable filters.

Burcea et al. propose a dedicated location matching engine in [7] that processes events matched by the filtering engine of a traditional attribute-based publish/subscribe system. If a producer or consumer is static, its geographic coordinates are directly encoded into the event or subscription as attributes. Otherwise, an identifier for the mobile client is added to outgoing events or subscriptions and the client periodically reports its current location. The location matching engine uses this identifier to retrieve the last known location of the respective client to perform location-based filtering. As location-based filtering is not part of the filter engine itself, this information cannot be utilized for optimizing filter-based routing of events in a distributed broker network.

Cugola et al. also separate traditional filtering from context matching in [17] by proposing an Application Programming Interface (API) to set a client's context when publishing or subscribing. The context is then distributed to other brokers via a dedicated protocol and routing entries are updated accordingly. Consequently, context information is decoupled from subscriptions, thereby reducing the overhead within the broker network. The authors state that their scheme performs well whenever the contextual filters are highly selective and rarely updated. While the former holds true within the scenario considered in this work, the latter does not apply given the dynamics of mobile clients.

Jayaram et al. propose parameterized subscriptions in [41]. Here, a client can use so-called *broker variables* within a subscription. Each broker variable corresponds to a time-dependent value. In case of location-based publish/subscribe, there could be two such variables: the current latitude and longitude of a mobile client. Clients issue update messages to their respective broker to update their broker variables. A similar concept is also proposed in [38]. The authors strive to maintain the scalability properties of a broker network by replacing costly re-subscriptions with more lightweight update procedures. We apply the concept of broker variables within our framework, specifically focusing on the transfer of such variables during mechanism transitions between filter schemes, as detailed in Sect. 4.2.

To deal with frequent location updates, dedicated subscription models and the corresponding filter schemes have been proposed in the literature [3, 5, 12, 22, 43, 56, 105]. In all these approaches, *location-based filtering* is a key primitive of the filter scheme, resulting in increased efficiency and expressiveness in dynamic scenarios.

An intuitive representation of location-based subscriptions is proposed by Brimicombe et al. in [5] with Space-Time Envelopes (STEs). STEs describe a circular area around a client's current location, as also proposed in [22]. However, an STE is further extended with a conical shape in the direction of the client's movement. The length of the cone is adapted based on the client's movement speed. All events that are published to a location that is covered by the STE are sent to the respective client. The technical realization of STEs is discussed in more detail in Section 4.2.4. Similar concepts to represent an (AoI) have been proposed for interest management in NVEs [58, 68]. Relying on STEs results in distinct subscriptions for each mobile client and involves rather complex calculations to match events against the geometric shape of the envelope. To reduce complexity, other schemes propose to subscribe clients to static areas instead.

Chen et al. rely on named locations to match events in [12]. A named location corresponds to an area described by a polygon or circle. Events can be published to specific named locations and subscriptions filter for events based on their target location. Given that the named locations are known in advance, the resulting filter scheme closely resembles generic channel-based filtering. The authors briefly motivate that mobile clients should locally detect their current channel based on their location to reduce the load on the broker. However, they do not further elaborate on how to achieve this goal. In [43], Jodlauk et al. propose a grid-based system. Clients are assigned to a cell within the grid based on their current location. By carefully choosing the size of the respective cells, the system can be adapted with respect to the desired accuracy of event dissemination and the resulting overhead.

Other designs restrict the representation of the AoI to a rectangular shape, referred to as Minimum Bounding Rectangle (MBR) [11, 54, 105, 107]. In [107], the authors focus on data structures for the representation of nested hierarchies. Here, an MBR can contain other MBRs to allow for fine-grained location-based information retrieval. The concept is extended and applied to publish/subscribe by Zhou et al. in [54, 105], now allowing overlapping MBRs. The authors propose an *R*-tree [30] based index structure for the resulting location-based subscriptions and devise an efficient filtering algorithm utilizing the proposed index structure. Chen et al. further discuss the utilization of a tree-based index structure for spatial top-k publish/subscribe in [11], also relying on the MBR to describe objects and regions of interest. In their work, instead of notifying a client of all matches, only the top k results according to a given cost metric are to be forwarded. Supporting top-k queries (or stateful operators in general [20, 49]) does not lie within the scope of our work. Still, it might be a promising technique to reduce transmission of less relevant events to mobile clients on a semantic level. Given the results of the performance analysis presented in [11, 54, 105, 107], it can be assumed that existing MBR-based schemes as utilized in our work can further benefit from more elaborate index structures. Research on location queries and databases for moving objects, as surveyed in [40], can further aid in realizing efficient filtering at a broker.

Holzer et al. introduce *context aspects* [34], arbitrarily complex filter expressions that operate on locally available context variables. The authors enable location-based filtering by performing the respective distance calculations within such an expression. However, their work is focused on the practical aspects of integrating context aspects into the EVENTJAVA [21] language.

Location-based mobile social applications require support for physical mobility *and* interest mobility, while at the same time exhibiting strong locality properties in the interaction between users. However, none of the aforementioned approaches supports both, physical mobility and interest mobility, and the consequences of a combined utilization are not yet studied. Additionally, the state of the art in exploiting locality properties in mobile scenarios focuses on pure MANETs, although they can also be beneficial in a hybrid environment relying on cloud-based brokers. Supporting physical mobility and interest mobility at the same time and in a locality-aware fashion requires an adaptive approach due to the dynamics of the scenario discussed in Sect. 2.1. Hence, we discuss existing approaches to adapt publish/subscribe systems to dynamic environments in the following.

3.2 Adaptivity in Publish/Subscribe Systems

In the following, we provide a broader perspective on techniques used to adapt publish/subscribe systems and discuss their applicability for the scenario of location-based mobile applications. Existing approaches for adaptivity in publish/subscribe can be categorized as *broker-centric* or *client-centric*. Broker-centric approaches focus on adaptation techniques applied within broker networks, usually transparent to clients. Client-centric approaches aim at utilizing the resources of potentially mobile clients during filtering and distribution of events.

3.2.1 Broker-Centric: Adaptivity in Distributed Broker Networks

Techniques such as subscription bundling [97] and interest clustering based on filter similarity [42, 50, 52, 65] have been proposed for distributed broker networks. These techniques are used to limit forwarding of events within a broker network by grouping clients with similar interest at brokers that are directly connected. Additionally, by bundling subscriptions and utilizing filter-based routing, events can be dropped early based on a broker's knowledge about the aggregated interest of each neighboring broker.

Li et al. propose a representation of attribute-based subscriptions as modified binary decision diagrams in [52], allowing a unified approach to subscription merging, covering, and routing. The authors further propose to utilize statistical information about the popularity of individual attributes and their logical relationships to adapt and optimize the routing procedure. A similar concept is also proposed by Mühl et al. in [65], significantly reducing the amount of control messages.

In their content-based publish/subscribe system BERETTA [42], Jayram et al. rely on interval-based normalization of subscriptions. During the subscription normalization, relational operators are transformed into expressions relying solely on the inclusion operator ($\in$). This is utilized in BERETTA's routing and matching algorithm to reduce the overhead associated with subscription updates. The authors state that their algorithm supports subscription summarization as proposed in [97].

Similar techniques have also been proposed in the field of P2P-based publish/subscribe systems [27, 73, 95]. Here, the publish/subscribe system does not provide a clear distinction between brokers and clients. Instead, a client contributes its resources to the P2P network [92]. In MAGNET [27], the core task is to construct efficient multicast groups for clients with similar interests. Therefore correlation patterns among subscriptions are utilized to group clients within an underlying Distributed Hash Table (DHT). Similarly, Terpstra et al. [95] rely on a modified version of the CHORD DHT [94] to realize content-based publish/subscribe in a P2P fashion. Besides DHTs, unstructured rendezvous-based approaches such as BUBBLESTORM [96] have been proposed as substrate for publish/subscribe systems. Similarly, the P2P publish/subscribe system RAPPEL [73] constructs and maintains its own *friendship overlay*, grouping clients with similar interest in the underlying message substrate.

Utilizing locality in interests to achieve locality in communication is a key design decision in most P2P-based publish/subscribe systems for fixed networks. However, as shown for other application areas of P2P systems, the dynamics caused by clients joining and leaving the system poses additional challenges and limits the efficiency of the aforementioned optimization approaches for broker networks [92]. This effect is even more severe if mobile clients are expected to contribute their resources and act as brokers, as previously discussed in Sect. 3.1.2.

Within our work, we focus on the communication between a broker and its associated mobile clients, specifically addressing the challenges that arise within a mobile application scenario. Given our focus on location-based applications, interest in content is directly tied to a client's physical location. It is therefore reasonable to assume that the aforementioned approaches to interest-based clustering would lead to an assignment based on the location of clients. As we utilize a generic attribute-based subscription model, a broker network could still benefit from the concepts and algorithms discussed in this section. Therefore, provider-centric approaches to adaptivity can complement the contributions presented in this thesis.

3.2.2 Client-Centric: Adaptivity Through Reconfigurable Middleware

In order to better adapt to the dynamics of mobile scenarios, several middleware solutions have been proposed [18, 28, 31, 32, 88–90] that alter the behavior of publish/subscribe systems by actively utilizing the capabilities of client devices. These capabilities are not limited to communication interfaces for ad hoc connectivity. Additional sensor input can be utilized to derive the client's current context and adapt the middleware configuration accordingly.

Holzer et al. propose ALPS [31, 32], a location-based publish/subscribe middleware for MANETs. The message dissemination in ALPS can be parameterized with respect to the intended coverage of a message, resembling the rendezvous-based approach BUBBLESTORM for fixed networks [96]. The respective parameters of the dissemination module are adapted based on the currently observed set of events and subscriptions in the neighborhood of a client. However, the approach is not evaluated for dynamic subscriptions based on a client's current location, and the required coordination among clients produces significant message overhead. Gokhale et al. also propose to adapt the transport mechanism [28]. They follow a closed-loop approach utilizing machine learning to derive suitable rules for an adaptation of the transport mechanism based on monitoring data. In contrast to our work, the approach is currently limited to the transport protocol. We adapt multiple functional aspects of our system through the generalized design concept of mechanism transitions.

Similar dissemination mechanisms have recently also been proposed for middleware platforms for ubiquitous systems, as surveyed by Bellavista et al. in [4] and Knappmeyer et al. in [47]. Based on their surveys, the authors explicitly highlight a field for future research. According to [4], systems need to "[…] self-adapt autonomously by dynamically combining most suitable data distribution methods and techniques […] depending on current management conditions". This motivates our research on mechanism transitions as an enabler for the envisioned dynamic combinations within a system.

With GREEN [88], Sivaharan et al. propose a middleware that is configured by composing pluggable components for different functional aspects. GREEN supports reconfigurability of components during runtime. However, the authors do not evaluate the performance impact during the reconfiguration. Instead, they focus on the memory overhead and loading time for components on the Windows CE mobile OS. The authors do not discuss a locally limited reconfiguration for a subset of clients, which constitutes an essential requirement in a heterogeneous scenario.

Cugola et al. study the concept of reconfiguration during runtime with REDS [18]. The authors design REDS such that additional mechanisms can be integrated and utilized in the middleware, focusing their work on the software engineering aspects. Within REDS, functionally equivalent mechanisms are hidden behind a unified interface. Such mechanisms include local dissemination of events via ad hoc networking protocols. From an architectural point of view, the representation of functional aspects of the system as pluggable components constitutes an important step towards enabling their exchange during runtime. However, the authors neither evaluate the systems' behavior during reconfigurations, nor the impact of client mobility.

Our concept of mechanism proxies discussed in Sect. 5.2.1 follows the same design goal of exposing functionality covered by multiple specific mechanisms via a unified interface. However, the *seamless* execution of transitions is a key requirement. Consequently, we propose mechanisms for state transfer and transformation, which we evaluate extensively. Additionally, we specifically address heterogeneous scenarios and enable the publish/subscribe system to adapt locally and for a confined set of clients based on application-specific mobility and workload characteristics. This is achieved by applying and extending the design concept of mechanism transitions (as introduced in Sect. 2.4). Therefore, we discuss approaches that apply transitions to other application domains in the following section.

3.3 Adaptivity Through Mechanism Transitions

The concept of transitions between mechanisms as introduced in Sect. 2.4 is studied within the Collaborative Research Centre "MAKI". The realization of transition-enabled communication systems and the applicability of generalized concepts for the execution of transitions has also been studied in other application domains. Within this section, we highlight key insights obtained from related work and our own collaborations on mechanism transitions in video streaming and network monitoring.

Based on our contributions on transitions in P2P video streaming systems [77], we explored the concept further within the live streaming systems TRANSIT [100] and TOPT [83]. In contrast to systems for scalable and adaptive Video on Demand (VoD) streaming [1, 75, 108], a stream is only available at one source and cannot be prefetched by clients. Consequently, timely dissemination of the most recent video segments is key to achieve high Quality of Experience (QoE) [100]. In contrast to the work conducted in this thesis, we modeled transitions as local adaptation of utilized mechanisms. Instead of coordinating transitions for the whole system or a subset of clients, each client individually decides which mechanisms to use based on local knowledge. To maintain compatibility among clients without requiring global coordination, available mechanisms for scheduling and neighborhood maintenance need to operate on common messages and data types. While this limits the set of mechanisms that could be integrated into the system, it prevents additional coordination overhead for the execution of transitions [100]. Subsequent contributions highlighted the achieved adaptivity for large-scale live video streaming scenarios [82]. These approaches utilized layer-encoded video [109] to adapt the bandwidth requirements of the streaming process to the current environmental conditions. The benefits of adapting scheduling mechanisms based on user perceived quality has been modeled and evaluated by Wang et al. in [98], further highlighting the issue of cross-layer dependencies between mechanisms such as TCP and the video streaming protocol DASH.

Within TRANSIT and its extension TOPT, individual mechanisms are already separated via interfaces. However, mechanisms are not actually switched. Instead, they operate in parallel at different levels of utilization. Scheduling in TRANSIT, for example, is realized by two basic mechanisms: a pull-based request mechanism and a push-based flow mechanism. Depending on its local buffer state and the availability of data within its neighborhood, a client utilizes both mechanisms at varying degrees. In contrast, within this thesis, we execute transitions by switching between two completely separated mechanisms, without any parallel utilization. Thereby, we provide a clear distinction between a transition and a self-adaptive mechanism.

Other approaches explore the utilization of transitions for streaming and uploading of user-generated content [45, 78, 101, 102]. In [101, 102], the encoding scheme of the uploaded stream is altered based on the expected perceived quality on the receiving end, freeing network resources as a side effect. In [78], nearby mobile clients' resources are utilized for collaborative uploading of video streams in heterogeneous mobile environments.

In addition to video streaming, the concept of mechanism transitions has been proposed and explored within the domain of mobile network monitoring [80, 81, 93]. The monitoring system CRATER [80] can switch between a centralized and a distributed collection mechanism. This is motivated by the need to lower the load on the cellular infrastructure caused by background services such as monitoring, especially in situations with a high client density. Additionally, Stingl et al. motivate transitions between distinct monitoring mechanisms in MANETs in [93]. In [80], the density of clients determines whether the system switches to a gradient-based routing mechanism utilizing local ad hoc connectivity, rather than reporting results to the central entity via the mobile network.

In [81], the concept of distributed data collection is taken one step further by the utilization of gateway selection algorithms. Selecting gateways in an intelligent way enables the monitoring system to offload traffic from the cellular network without loosing completeness or reducing the accuracy. Additionally, specific fairness characteristics can be enforced to raise acceptance among clients that act as gateways. The framework of gateway selection algorithms developed in [81] is utilized in our work, as later discussed in Sect. 4.3.4.

In addition to the aforementioned examples, transitions have been studied in the context of topology adaptation for wireless sensor networks. In [91], the authors propose transitions that alter the structure of a connectivity graph by adding and removing edges representing potential connections between sensor nodes. By removing long edges in a process referred to as *topology control* while still maintaining connectivity, the energy used to transmit data can be reduced. The concept of transitions to add and remove edges in a connectivity graph enables the generic construction of incremental topology control algorithms, as proposed in [46].

3.4 Summary and Identified Research Gap

We previously discussed related work with respect to three questions motivated by our scenario. In the following, we briefly summarize key findings for each question and highlight the research gap addressed in our work.

How to support client, broker, and interest mobility in publish/subscribe?

Individually, each aspect of mobility has been studied extensively, as discussed in Sect. 3.1. Approaches dealing with client mobility focus on handling intermittent connection loss, enabling client handovers between brokers, and ensuring delivery of events during handover. Related work on broker mobility is limited to MANETs, where a subset of the mobile clients acts as brokers. The goal of disseminating events to all interested clients is tackled by structured and unstructured approaches, each with its respective advantages as discussed. In addition to physical mobility, support for interest mobility is a key requirement for location-based applications. This can be achieved through subscription models and filter schemes that support location-based filtering. In our work, we study the combination of mechanisms for location-based filtering and mechanisms for locality-aware event dissemination that tackle broker and client mobility. Furthermore, we propose mechanism combinations that offload the cellular connection at virtually no additional cost by benefiting from the filtering process of a publish/subscribe system.

How to achieve adaptivity in dynamic and heterogeneous environments?

As discussed in Sect. 3.2, most contributions addressing adaptive publish/subscribe systems focus on filtering and routing within the broker network. To this end, filter-based routing, subscription summarization, and filter merging allow the system to adapt to dynamic workloads. However, these techniques only benefit the broker network and do not address mobile clients. To this end, a reconfigurable middleware for publish/subscribe in MANETs aims to incorporate a mobile client's context and capabilities. Existing approaches reconfigure the middleware during runtime to adapt to dynamic environments. However, the impact of such reconfigurations on the system's performance was not studied so far. Furthermore, reconfigurations always affect all clients within the respective network. This limits the adaptivity with respect to heterogeneous client dynamics arising in mobile social applications. In our work, we propose adaptation based on application-specific attraction points for a regionally confined group of clients, focusing explicitly on the *last mile* between broker and mobile client. We study the adaptivity of publish/subscribe systems during runtime, focusing on the impact of our proposed mechanisms on the performance and cost characteristics of the overall system.

How are mechanism transitions utilized to adapt communication systems?

Adapting systems by executing transitions between individual mechanisms proved promising for other application domains, e.g., video streaming and network monitoring. Mechanism transitions enable the utilization of state of the art mechanisms previously reviewed in this chapter. Therefore, applying the design concept of mechanism transitions to publish/subscribe is a promising approach towards achieving adaptivity under dynamic conditions. Additionally, it allows us to utilize and combine mechanisms for location-based filtering and mechanisms for locality-aware event dissemination, going beyond the capabilities of the previously discussed middleware solutions. We extend the concept of transitions to execute mechanism transitions for a subset of a broker's clients, addressing the heterogeneity of conditions in mobile social applications. We further propose and evaluate mechanisms for seamless state transfer that maintain a constantly high performance during transitions.

In this chapter, we identified and surveyed mechanisms that enable (i) location-based filtering or locality-aware dissemination and (ii) adaptivity to dynamic environmental conditions. However, the state of the art does not *combine* mechanisms for location-based filtering and locality-aware dissemination, which is an essential step for efficient communication in location-based mobile social applications. Furthermore, adaptivity is either limited to broker networks or cannot deal with heterogeneous application-specific mobility characteristics. In our work, we design seamless transitions between mechanisms for location-based and locality-aware publish/subscribe, addressing the requirement of fine-grained adaptivity to heterogeneous conditions arising from location-based mobile social applications. As motivated, we address the design of transitions in a mechanism-specific manner in BYPASS.KOM, presented in Chap. 4, and generalize our findings in the SIMONSTRATOR.KOM platform discussed in Chap. 5.

References

1. Abboud O (2012) Quality adaptation in peer-to-peer video streaming: supporting heterogeneity and enhancing performance using scalable video coding. PhD thesis. TU Darmstadt
2. Arnold K, Scheifler R, Waldo J, O'Sullivan B, Wollrath A (1999) Jini specification. Addison-Wesley Longman Publishing Co., Inc
3. Baldoni R, Marchetti C, Virgillito A, Vitenberg R (2005) Content-based publish-subscribe over structured overlay networks. In: Proceedings IEEE international conference on distributed computing systems (ICDCS). IEEE, pp 437–446
4. Bellavista P, Corradi A, Fanelli M, Foschini L (2012) A survey of context data distribution for mobile ubiquitous systems. ACM Comput Surv (CSUR) 44(4):24
5. Brimicombe A, Li Y (2006) Mobile space-time envelopes for location-based services. Trans GIS 10(1):5–23
6. Broder A, Mitzenmacher M (2004) Network applications of bloom filters: a survey. Int Math 1(4):485–509
7. Burcea I, Jacobsen H-A (2003) L-ToPSS-push-oriented locationbased services. In: International workshop on technologies for E-services. Springer, pp 131–142
8. Claudia C, Antonella M, Scopigno R (2015) Vehicular ad hoc networks: standards, solutions, and research. Springer
9. Caporuscio M, Carzaniga A, LWolf A (2003) Design and evaluation of a support service for mobile, wireless publish/subscribe applications. IEEE Trans Softw Eng 29(12):1059–1071
10. Carzaniga A, Rosenblum DS, Wolf AL (2001) Design and evaluation of a wide-area event notification service. ACM Trans Comput Syst (TOCS) 19(3):332–383
11. Chen L, Cong G, Cao X, Tan K-L (2015) Temporal spatialkeyword top-k publish/subscribe. In: Proceedings international conference on data engineering. IEEE, pp 255–266
12. Chen X, Chen Y, Rao F (2003) An efficient spatial publish/- subscribe system for intelligent location-based services. In: Proceedings international workshop on distributed event-based systems (DEBS). ACM, pp 1–6
13. Cheung AKY, Jacobsen H-A (2010) Publisher placement algorithms in content-based publish/subscribe. In: Proceedings international conference on distributed computing systems (ICDCS). IEEE, pp 653–664
14. Cilia M, Fiege L, Haul C, Zeidler A, Buchmann AP (2003) Looking into the past: enhancing mobile publish/sub scribe middleware. In: Proceedings international workshop on distributed event based systems (DEBS). ACM, pp 1–8

15. Costa P, Picco GP (2005) Semi-probabilistic content-based publish subscribe. In: Proceedings IEEE international conference on distributed computing systems (ICDCS). IEEE, pp 575–585
16. Costa P, Mascolo C, Musolesi M, Picco GP (2008) Socially-aware routing for publish-subscribe in delay-tolerant mobile ad hoc networks. IEEE J Sel Areas Commun 26(5):748–760
17. Cugola G, Margara A, Migliavacca M (2009) Contextaware publish-subscribe: model, implementation, and evaluation. In: Proceedings IEEE symposium on computers and communications (ISCC). IEEE, pp 875–881
18. Cugola G, Picco GP (2006) REDS: a reconfigurable dispatching system. In: Proceedings international workshop on software engineering and middleware (SEM). ACM, pp 9–16
19. Datta A, Quarteroni S, Aberer K (2004) Autonomous gossiping: a self-organizing epidemic algorithm for selective information dissemination in wireless mobile ad-hoc networks. In: Semantics of a networked world. Semantics for Grid Databases. Springer, pp 126–143
20. Demers A, Gehrke J, Hong M, Riedewald M, White W (2006) Towards expressive publish/subscribe systems. In: International conference on extending database technology. Springer, pp 627–644
21. Eugster P, Jayaram KR (2009) EventJava: an extension of Java for event correlation. In: European conference on object-oriented programming. Springer, pp 570–594
22. Eugster PTh, Garbinato B, Holzer A (2005) Location-based publish/subscribe. In: Proceedings IEEE international symposium on network computing and applications. IEEE, pp 279–282
23. Fiege L, Gärtner FC, Kasten O, Zeidler A (2003) Supporting mobility in content-based publish/subscribe middleware. Proceedings ACM/IFIP/USENIX 2003 international conference on middleware. Springer-Verlag, New York Inc, pp 103–122
24. Frey D, Roman G-C (2007) Context-aware publish subscribe in mobile ad hoc networks. In: Proceedings international conference on coordination languages and models. Springer, pp 37–55
25. Friedman R, Shulman AK (2013) A density-driven publish subscribe service for mobile ad-hoc networks. Ad Hoc Networks 11(1):522–540
26. Gaddah A, Kunz T (2010) Extending mobility to publish/subscribe systems using a pro-active caching approach. Mobile Inf Syst 6(4):293–324
27. Girdzijauskas S, Chockler G, Vigfusson Y, Tock Y, Melamed R (2010) Magnet: practical subscription clustering for internet-scale publish/subscribe. In: Proceedings ACM international conference on distributed event-based systems (DEBS). ACM, pp 172–183
28. Gokhale A, Schmidt DC, Hoffert J (2011) Timely autonomic adaptation of publish/subscribe middleware in dynamic environments. Int J Adapt Resilient Auton Syst. 2.4, pp 1–24. https://doi.org/10.4018/jaras.2011100101
29. Groß C, Stingl D, Gottron C, Richerzhagen B, Münker C, Hausheer D (2012) Harnessing Mobile Ad Hoc Communication for Decentralized Location-Based Services. Technical report Peer-to- Peer Systems Engineering Lab, TU Darmstadt, Germany
30. Guttman A (1984) R-trees: a dynamic index structure for spatial searching. vol 14. 2. ACM
31. Holzer A, Eugster P, Garbinato B (2013) Evaluating implementation strategies for location-based multicast addressing. IEEE Trans Mobile Comput 12(5):855–867
32. Holzer A, Eugster P, Garbinato B (2012) Alps-adaptive location-based publish/subscribe. Computer networks 56(12):2949–2962
33. Holzer A, Vessaz F, Pierre S, Garbinato B (2011) PLAN-B: proximity-based lightweight adaptive network broadcasting. In: Proceedings international symposium on network computing and applications (NCA). IEEE, pp 265–270
34. Holzer A, Ziarek L, Jayaram KR, Eugster P (2012) Abstracting context in event-based software. In: Transactions on aspect-oriented software development IX. Springer, pp 23–167
35. Hu S, Muthusamy V, Li G, Jacobsen H-A (2009) Transactional mobility in distributed content-based publish/subscribe systems. In: Proceedings IEEE international conference on distributed computing systems (ICDCS). IEEE, pp 101–110
36. Huang Y, Garcia-Molina H (2001) Publish/subscribe in a mobile enviroment. In: Proceedings ACM international workshop on data engineering for wireless and mobile access. ACM, pp 27–34

37. Huang Y, Garcia-Molina H (2003) Publish/subscribe tree construction in wireless ad-hoc networks. In: Proceedings international conference on mobile data management. Springer, pp 122–140
38. Huang Y, Garcia-Molina H (2007) Parameterized subscriptions in publish/subscribe systems. Data Knowl Eng 60(3):435–450
39. Hunkeler U, Truong HL, Stanford-Clark A (2008) MQTT-S - a publish/subscribe protocol for wireless sensor networks. In: Proceedings international conference on communication systems software and middleware and workshops (comsware). IEEE, pp 791–798
40. Ilarri S, Mena E, Illarramendi A (2010) Location-dependent query processing: where we are and where we are heading. ACM Comput Surv (CSUR) 42(3):12
41. Jayaram KR, Eugster P, Jayalath C (2013) Parametric contentbased publish/subscribe. ACM Trans Comput Syst (TOCS) 31(2):4
42. Jayaram KR, Wang W, Eugster P (2015) Subscription normalization for effective content-based messaging. IEEE Transactions on parallel and distributed systems (TPDS) 26(11):3184–3193
43. Jodlauk G, Rembarz R, Xu Z (2011) An optimized gridbased geocasting method for cellular mobile networks. In: Proceedings ITS world congress
44. Khelil A, Pedro José M, Christian B, Kurt R (2007) Hypergossiping: a generalized broadcast strategy for mobile ad hoc networks.. Ad Hoc Networks 5(5):531–546
45. Khemmarat S, Renjie Z, Krishnappa DK, Gao L, Zink M (2012) Watching user generated videos with prefetching. Signal Process Image Commun 27(4):343–359
46. Kluge R, Stein M, Varró Andy Schürr G, Hollick M, Mühlhäuser M (2017) A systematic approach to constructing families of incremental topology control algorithms using graph transformation. In: Software & systems modeling, pp 1–41
47. Knappmeyer M, Kiani SL, Reetz ES, Baker N, Tonjes R (2013) Survey of context provisioning middleware. IEEE Commun. Surv Tutorials 15(3):1492–1519
48. Ko Y-B, Vaidya NH (1999) Geocasting in mobile ad hoc networks: Location-based multicast algorithms. In: Proceedings IEEE workshop on mobile computing systems and applications (WMCSA). IEEE, pp 101–110
49. Koldehofe B, Ottenwälder B, Rothermel K, Ramachandran U (2012) Moving range queries in distributed complex event processing. In: Proceedings ACM international conference on distributed event-based systems (DEBS). ACM, pp 201–212
50. Lehn M, Rehner R, Buchmann A (2013) Distributed optimization of event dissemination exploiting interest clustering. In: Proceedings IEEE conference on local computer networks (LCN). IEEE, pp 328–331
51. Leontiadis I, Costa P, Mascolo C (2009) A hybrid approach for content-based publish/subscribe in vehicular networks. Pervasive Mobile Comput 5(6):697–713
52. Li G, Hou S, Jacobsen H-A (2005) A unified approach to routing, covering and merging in publish/subscribe systems based on modified binary decision diagrams. In: Proceedings IEEE international conference on distributed computing systems (ICDCS). IEEE, pp 447–457
53. Li G, Muthusamy V, Jacobsen H-A (2008) Adaptive contentbased routing in general overlay topologies. Proceedings ACM/IFIP/USENIX international conference on middleware. Springer-Verlag, New York Inc, pp 1–21
54. Li G, Wang Y, Wang T, Feng J (2013) Location-aware publish/subscribe. In: Proceedings ACM SIGKDD international conference on knowledge discovery and data mining. ACM, pp 802–810
55. Li J, Blake C, SJ De Couto D, Lee HI, Morris R (2001) Capacity of ad hoc wireless networks. In: Proceedings international conference on mobile computing and networking. ACM, pp 61–69
56. Liang S, Gao Y et al (2010) Real-time notification and improved situational awareness in fire emergencies using geospatial-based publish/subscribe. Int J Appl Earth Observ Geoinf 12(6):431–438
57. Lin CR, Gerla M (1997) Adaptive clustering for mobile wireless networks. IEEE J Sel Areas Commun 15(7):1265–1275

58. Liu ES, Theodoropoulos GK (2014) Interest management for distributed virtual environments: A survey. ACM computing surveys (CSUR) 46(4):51
59. Maihofer C (2004) A survey of geocast routing protocols. IEEE communications surveys & tutorials 6:2
60. Meier R, Cahill V (2002) Steam: event-based middleware for wireless ad hoc networks. In: Proceedings international conference on distributed computing systems workshops. IEEE, pp 639–644
61. Meier R, Cahill V (2010) On event-based middleware for locationaware mobile applications. IEEE Trans Softw Eng 36(3):409–430
62. Mitra P, Poellabauer C (2012) Efficient group communications in location aware mobile ad-hoc networks. Pervasive Mobile Comput 8(2):229–248
63. Moon S-C, Ko Y, Lee D (2007) A fast path recovery scheme for publish/subscribe in mobile ad hoc networks. In: Proceedings IEEE conference on computer and information technology (CIT). IEEE, pp 435–440
64. Mottola L, Cugola G, Pietro PG (2008) A self-repairing tree topology enabling content-based routing in mobile ad hoc networks. IEEE Trans Mobile Comput 7(8):946–960
65. Mühl G, Fiege L, Buchmann A (2002) Filter similarities in content-based publish/subscribe systems. In: Trends in network and pervasive computing—ARCS 2002. Springer, pp 224–238
66. Muhl G, Ulbrich A, Herrman K (2004) Disseminating information to mobile clients using publish-subscribe. IEEE Int Comput 8(3):46–53
67. Muhl G, Fiege L, Gartner FC, Buchmann A (2002) Evaluating advanced routing algorithms for content-based publish/subscribe systems. In: Proceedings IEEE international symposium on modeling analysis and simulation of computer and telecommunications systems (MASCOTS). IEEE, pp 167–176
68. Najaran MT, Hu S-Y, Hutchinson NC (2014) SPEX: scalable spatial publish/subscribe for distributed virtual worlds without borders. In: Proceedings ACM multimedia systems conference (MMSys). ACM, pp 127–138
69. Navas JC, Imielinski T (1997) GeoCast—geographic addressing and routing. In: Proceedings ACM/IEEE international conference on mobile computing and networking. ACM, pp 66–76
70. Newton B, Aikat J, Jeffay K (2016) Geographic routing in extreme- scale highly-dynamic mobile ad hoc networks. In: Proceedings IEEE international symposium on modeling, analysis and simulation of computer and telecommunication systems (MASCOTS). IEEE, pp 205–210
71. Ottenwälder B, Koldehofe B, Rothermel K, Ramachandran U (2013) Migcep: operator migration for mobility driven distributed complex event processing. In: Proceedings ACM international conference on distributed event-based systems (DEBS). ACM, pp 183–194
72. Ottenwälder B, Koldehofe B, Rothermel K, Hong K, Lillethun D, Ramachandran U (2014) Mcep: a mobility-aware complex event processing system. ACM Transactions on internet technology (TOIT) 14(1):6
73. Patel JA, Rivière É, Gupta I, Kermarrec A-M (2009) Rappel: exploiting interest and network locality to improve fairness in publish-subscribe systems. Comput Netw 53(13):2304–2320
74. Pietzuch PR, Bacon JM (2002) Hermes: a distributed event-based middleware architecture. In: Proceedings conference on distributed computing systems workshops. IEEE, pp 611–618
75. Pussep K (2011) Peer-assisted video-on-demand: cost reduction and performance enhancement for users, overlay providers, and network operators. PhD thesis. TU Darmstadt
76. Rebecchi F, Amorim Marcelo DD, Conan V, Passarella A, Bruno R, Conti M (2015) Data offloading techniques in cellular networks: a survey. IEEE Commun Surv Tutorials 17(2):580–603
77. Richerzhagen B (2012) Supporting transitions in peer-to-peer video streaming. In: Master's thesis, Technische Universität Darmstadt
78. Richerzhagen B, Wulfheide J, Koeppl H, Mauthe A, Nahrstedt K, Steinmetz R (2016) Enabling crowdsourced live event coverage with adaptive collaborative upload strategies. In: Proceedings17th international symposium on a world of wireless, mobile and multimedia networks (WoWMoM). IEEE, pp 1–3

79. Richerzhagen B, Richerzhagen N, Schönherr S, Hark R, Steinmetz R (2016) Stateless gateways—reducing cellular traffic for event distribution in mobile social applications. In: Proceedings 25th interna tional conference on computer communication and networks (ICCCN). IEEE, pp 1–9
80. Richerzhagen N, Stingl D, Richerzhagen B, Mauthe A, Steinmetz R (2015) Adaptive monitoring for mobile networks in challenging environments. In: Proceedings 24th international conference on computer communication and networks (ICCCN). IEEE, pp 1–8
81. Richerzhagen N, Richerzhagen B, Hark R, Stingl D, Steinmetz R (2016) Limiting the footprint of monitoring in dynamic scenarios through multi-dimensional offloading. In: Proceedings 25th international conference on computer communication and networks (ICCCN). IEEE, pp 1–9
82. Rückert J (2016) Large-scale live video streaming over the internet - efficient and flexible content delivery using network and application-layer mechanisms. PhD thesis. TU Darmstadt
83. Rückert J, Richerzhagen B, Lidanski E, Steinmetz R, Hausheer D (2015) TopT: supporting flash crowd events in hybrid overlay-based live streaming. In: Proceedings 14th IFIP networking conference (Networking). IEEE, pp 1–9
84. Ruiz P, Bouvry P (2015) Survey on broadcast algorithms for mobile ad hoc networks. ACM Computing Surveys (CSUR) 48(1):8
85. Saurez E, Hong K, Lillethun D, Ramachandran U, Ottenwälder B (2016) Incremental deployment and migration of geodistributed situation awareness applications in the fog. In: Proceedings ACM international conference on distributed and event-based systems (DEBS). ACM, pp 258–269
86. Segall B, Arnold D(1997) Elvin has left the building: a publish/subscribe notification service with quenching. In: Proceedings of AUUG. Brisbane, Australia, pp 243–255
87. Siegel J, Frantz D (2000) CORBA 3 fundamentals and programming, vol 2. Wiley, New York, NY, USA
88. Sivaharan T, Blair G, Coulson G (2005) Green: a configurable and re-configurable publish-subscribe middleware for pervasive computing. In: Proceedings international conferences on the move to meaningful internet systems (OTM). Springer, pp 732–749
89. Sørensen C-F, Wu M, Sivaharan T, Blair GS, Okanda P, Friday A, Duran-Limon H (2004) A context-aware middleware for applications in mobile ad hoc environments. In: Proceedings workshop on middleware for pervasive and ad-hoc computing (MPAC). ACM, pp 107–110
90. Souto E, Guimarães G, Vasconcelos G, Vieira M, Rosa N, Ferraz C, Kelner J (2006) Mires: a publish/subscribe middleware for sensor networks. Personal Ubiquitous Comput 10(1):37–44
91. Stein M, Frömmgen A, Kluge R, Löffler F, Schürr A, Buchmann A, Mühlhäuser M (2016) TARL: modeling topology adaptations for networking applications. In: Proceedings international symposium on software engineering for adaptive and self-managing systems (SEAMS). ACM, pp 57–63
92. Steinmetz R, Wehrle K (2005) What Is this "Peer-to-Peer" about? In: Peer-to-Peer systems and applications. Springer, pp 9–16
93. Stingl D (2014) Decentralized monitoring in mobile ad hoc networks—provisioning of accurate and location-aware monitoring information. PhD thesis. TU Darmstadt
94. Stoica I, Morris R, Karger D, Kaashoek MF, Balakrishnan H (2001) Chord: a scalable peer-to-peer lookup service for internet applications. ACM SIGCOMM computer communication review 31(4):149–160
95. Terpstra WW, Behnel S, Fiege L, Zeidler A, Buchmann AP (2003) A peer-to-peer approach to content-based publish/- subscribe. In: Proceedings international workshop on distributed event-based systems (DEBS). ACM, pp 1–8
96. Terpstra WW, Kangasharju J, Leng C, Buchmann AP (2007) Bubblestorm: resilient, probabilistic, and exhaustive peer-topeer search. In: ACM SIGCOMM computer communication review. vol 37. 4. ACM, pp 49–60
97. Triantafillou P, Economides A (2004) Subscription summarization: a new paradigm for efficient publish/subscribe systems. In: Proceedings international conference on distributed computing systems (ICDCS). IEEE, pp 562–571

98. Wang C, Rizk A, Zink M (2016) Squad: a spectrum-based quality adaptation for dynamic adaptive streaming over HTTP. In: Proceedings international conference on multimedia systems (MMSys). ACM, p 1
99. Wang H, Zhu T, Lei S, Yang W, Wang Y (2015) CMTG: a content-based mobile tendency geocast routing protocol in urban vehicular networks. Int J Distrib Sensor Netw (2015)
100. Wichtlhuber M, Richerzhagen B, Rückert J, Hausheer D (2014) TRANSIT: supporting transitions in peer-to-peer live video streaming. In: Proceedings IFIP networking conference (IFIP Networking). IEEE, pp 1–9
101. Wilk S (2016) Quality-aware content adaptation in digital video streaming. PhD thesis. TU Darmstadt
102. Wilk S, Zimmermann R, Effelsberg W (2016) Leveraging transitions for the upload of user-generated mobile video. In: Proceedings international workshop on mobile video. ACM, p 5
103. Yoneki E, Bacon J (2004) An adaptive approach to content-based subscription in mobile ad hoc networks. In: Percom Workshops. Citeseer, pp 92–97
104. Yoo S, Son JH, Ho Kim M (2009) A scalable publish/- subscribe system for large mobile ad hoc networks. J Syst Softw 82(7):1152–1162
105. Yu M, Guoliang L, Wang T, Feng J, Gong Z (2015) Efficient filtering algorithms for location-aware publish/subscribe. IEEE Trans Knowl Data Eng 27(4):950–963
106. Zeidler A (2005) A distributed publish/subscribe notification service for pervasive environments. PhD thesis. TU Darmstadt
107. Zhou Y, Xie X, Wang C, Gong Y, Ma W-Y (2005) Hybrid index structures for location-based web search. In: Proceedings ACM international conference on information and knowledge management. ACM, pp 155–162
108. Zink M (2013) Scalable video on demand: adaptive internet-based distribution. Wiley
109. Zink M, Schmitt J, Ralf S (2005) Layer-encoded video in scalable adaptive streaming. IEEE Trans Multimed 7(1):75–84

Chapter 4
BYPASS.KOM: Transitions in Event Brokering

In our analysis of related work, we identified a gap in the combined utilization of mechanisms for location-based filtering and locality-aware dissemination of events. As discussed in Chap. 3, applying these mechanisms to the dynamic scenario of location-based mobile social applications is further hindered by their limited adaptability. We designed BYPASS.KOM [19] as a framework to study the potential of mechanism transitions for location-based filtering and locality-aware event brokering. BYPASS.KOM enables us to include a wide range of existing mechanisms identified in our literature survey when studying the impact of transitions. At the same time, it allows us to combine mechanisms that alter the structure of the publish/subscribe system with mechanisms that alter the content of publications and subscriptions. Enabling this combined utilization of mechanisms is essential to address both, location-based filtering and locality-aware event dissemination.

In the following, we provide a brief conceptual overview of BYPASS.KOM. In Sect. 4.2, we present transitions between filter schemes as a promising methodology to support adaptive location-based filtering. In Sect. 4.3 we focus on transitions between mechanisms for locality-aware event dissemination, including ad hoc dissemination mechanisms and their utilization in hybrid, cloud-based scenarios. We further discuss the potential of transition-enabled gateway selection algorithms within such hybrid scenarios. The coexistence of mechanisms for location-based filtering and locality-aware event dissemination is finally discussed in Sect. 4.4. BYPASS.KOM plays a vital role in understanding the implications of mechanism transitions in a publish/subscribe system in the dynamic scenario of location-based mobile social applications. Although the mechanisms and transitions in BYPASS.KOM are specific to the domain of publish/subscribe, their understanding enables us to derive concepts for the design and evaluation of transition-enabled systems regardless of the application domain. These generalized concepts are detailed in Chap. 5 as part of the SIMONSTRATOR.KOM platform, which forms the basis for the evaluation of BYPASS.KOM presented in Chap. 6.

B. Richerzhagen, *Mechanism Transitions in Publish/Subscribe Systems*,
Springer Theses, https://doi.org/10.1007/978-3-319-92570-7_4

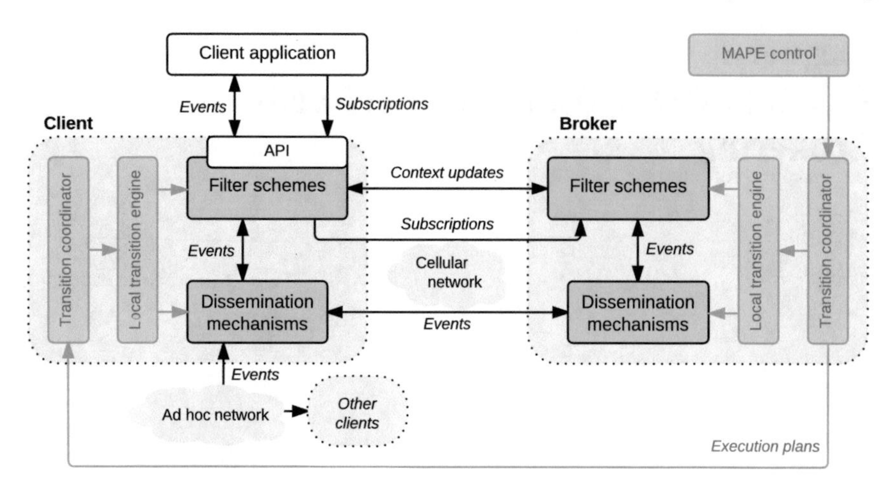

Fig. 4.1 Overview of BYPASS.KOM and its components

4.1 Conceptual Overview

BYPASS.KOM provides generic abstractions for two distinct types of mechanisms: *filter schemes* for location-based publish/subscribe and locality-aware *dissemination mechanisms*. This is illustrated in Fig. 4.1, with highlighted components being discussed in this chapter. A client application interacts with BYPASS.KOM through an API for location-based publish/subscribe, allowing the application to subscribe to events based on the client's current location. This API and the resulting structure of subscriptions issued by client applications is further discussed in Sect. 4.2. The API call is then processed by the client component of a filter scheme, as highlighted in the figure. This step might involve registering for location updates with the client's mobile OS. The subscription is then forwarded to the broker via the cellular network. Context associated with the subscription, e. g., the current location of a client, is updated according to the respective filter scheme via a custom context update protocol.

The broker uses stored subscriptions and the associated context information to filter incoming events. If an event matches a client's subscription, the event is sent to the client via a dissemination mechanism. The dissemination mechanisms are later discussed in detail in Sect. 4.3. A client receiving an event notifies the subscribed client application. Depending on the utilized dissemination mechanisms, it might further forward the event to other nearby clients via an ad hoc network. Similarly, events produced by a client are handled by a dissemination mechanism. Depending on the mechanism, the event is distributed via an ad hoc dissemination mechanism or sent to the cloud-based broker.

Filter schemes and event dissemination mechanisms are designed as *transition-enabled* mechanisms, allowing seamless transitions from one mechanism to another. In this chapter, we propose a methodology for the design of transition-enabled mechanisms and apply it to filter schemes and event dissemination mechanisms. Figure 4.1 includes additional components for the coordination and execution of transitions, such as the *transition coordinator* and a *local transition engine*, shown in gray. These components constitute a generalization of the concepts discussed in this chapter, realized in the SIMONSTRATOR.KOM platform. They are discussed in detail in Chap. 5.

Within BYPASS.KOM, we define three distinct types of transitions. A *total transition* is used to completely switch from one mechanism to another, for example to switch to a different filter scheme. This is modeled and discussed in detail in Sect. 4.2.3. However, if the filter scheme is only to be switched for a subset of subscribers, the broker needs to maintain parallel operation of multiple filter schemes to still ensure proper operation. The respective transition is termed a *partial transition*, described in detail in Sect. 4.2.3. This is an essential functionality to adapt to local dynamics and spatially distributed heterogeneities arising in the scenario of location-based mobile social applications. Lastly, we propose a unified approach to parameter reconfiguration or state modification of a mechanism by modeling the respective action as *self-transitions*. A self-transition does not switch between mechanisms, but instead alters state (including configuration parameters) of the current mechanism, as later discussed in Sect. 4.2.5.

4.2 Transitions Between Location-Based Filter Schemes

A filter scheme describes how events are matched against subscriptions. It therefore defines a model containing the structure of subscriptions and the available filter operations. For traditional publish/subscribe systems with static subscriptions, the resulting tasks for clients and brokers are clearly separated. Clients publish events and subscribe to information according to the expressiveness provided by the subscription model. Subscriptions and events are forwarded to the broker, which then determines the set of interested subscribers. These subscribers are then notified of the respective event.

For location-based filter schemes the clear separation between client and broker no longer holds, as up-to-date information about the locations of clients are required to determine the subscribers to an incoming event. Instead, an additional protocol between client and broker is required to update state information associated with the subscription. This state information, e. g., the current locations of subscribers, needs to be maintained by the broker alongside the set of subscriptions. Consequently, a filter scheme for location-based publish subscribe additionally consists of the components highlighted in Fig. 4.2 to provide means to update and maintain client context at the broker and utilize that context information to filter events. Note, that these additional components are not required if the location information is already part of the subscription model. This is the case for some early approaches to location-based publish/subscribe that extend the content-based subscription model. However,

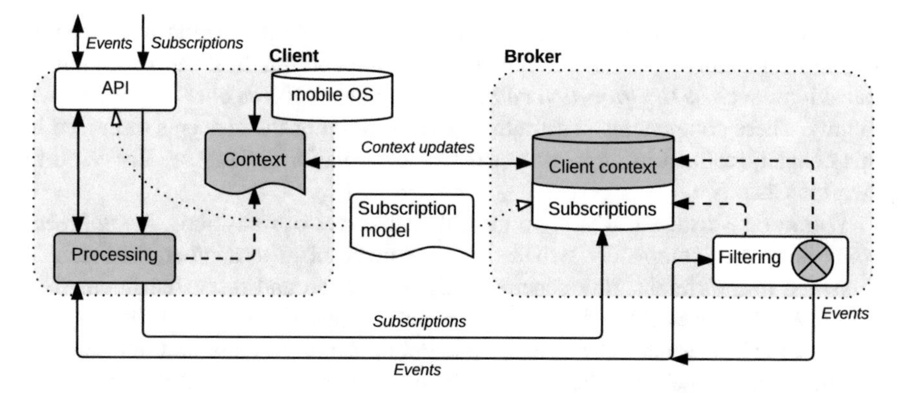

Fig. 4.2 Location-based filter schemes relying on a custom context update protocol

as discussed in Sect. 3.1.3, a decoupling of the subscription model and associated contextual information is beneficial and desired, especially in dynamic scenarios.

Regarding our goal of seamless transitions between filter schemes, a common subscription model for all filter schemes is furthermore required to be able to decouple the application logic from the filter scheme that is currently utilized. In the following, we present the API and underlying common subscription model of BYPASS.KOM as a foundation required to encapsulate filter schemes and, finally, execute transitions between different schemes.

4.2.1 Model and API for Location-Based Subscriptions

Regardless of the filter scheme, a common Application Programming Interface(API) for location-based publish/subscribe needs to be defined and provided to the application. To maintain compatibility with traditional attribute-based publish/subscribe systems, we extend the L1-API proposed by Pietzuch et al. [18] with specific methods for location-based publish/subscribe as proposed by Eugster et al. [4]. Instead of encoding information required for location-based publish/subscribe (i. e., the current location of a user and the size of the area of interest) within the subscription, we provide explicit methods for location-based subscriptions and events. Thereby, the context (location) is decoupled from the static content of the subscription (i. e., additional attributes) instead of being part of the subscription model.[1] According to Cugola et al. [3], this leads to increased efficiency and extensibility of the underlying

[1]The `subscribe` method proposed by Pietzuch et al. relies on `filter_expr`, a "[...] filter expression in any filtering language [...]" [18]. The subscription model is, thus, not defined as part of their L1-API. However, in their L3-API, the authors define an XML-based data model that relies on XPath queries for filtering. Using this model for location-based publish/subscribe would enforce tight coupling between context and static content of a subscription.

communication model. The resulting API for location-based publish/subscribe and the subscription model used in BYPASS.KOM are presented in the following.[2]

4.2.1.1 Subscription Model

The subscription model includes *events*, *operators*, and *filters* as defined in the following. Due to the decoupled handling of context information for location-based subscriptions, the subscription model proposed in this thesis does *not* include any additions for location-based publish/subscribe.

An event e consists of attribute-value tuples $\{(a_1, \texttt{value}_1), \ldots, (a_n, \texttt{value}_n)\}$ and additional application payload. We use $e_i.a$ to refer to the attribute a_i, and $e_i.\texttt{value}$ to refer to the associated value $\texttt{value}_i$ within the i-th tuple. Additionally, $e.a$ is used to refer to the set of all attributes $\{a_1, \ldots, a_n\}$ contained in the event. An application creates attributes through the API. Thereby, the representation of an attribute used within the publish/subscribe system is decoupled from the logic of the client application. Consequently, the full set of available attributes A is defined by the application.

An individual attribute $a \in A$ consists of a tuple $(\texttt{name}, \texttt{type})$ of the name and the data type of the attribute. To ease readability and prevent potential ambiguity for the application programmer, we require names of attributes to be unique regardless of the associated type. Therefore, the type of the attribute is sometimes omitted for readability, and the attribute is simply referred to by its name. As introduced for events, the name and type of an attribute is referred to using $a.\texttt{name}$ and $a.\texttt{type}$, respectively. Attributes can be instantiated with values of the respective type as a tuple: $(a, \texttt{value})$. As a shorthand, the notation $(a.\texttt{name}, \texttt{value})$ is used hereafter. An integer attribute "level" with a value of 42 is, thus, referred to as $((\text{"level"}, \mathbb{I}), 42)$, or, using the shorthand notation: ("level", 42). BYPASS.KOM supports numerical (integer and floating point) and string-based attributes and the corresponding operators, as discussed in the following. Additional types can be added by the application, given that the corresponding operators are also provided.

An operator $\texttt{op}$ identifies the function (or expression) that is to be executed by the broker when filtering events. The implementation and availability of a given operation depends on the type of the attribute. In contrast to attributes, operators with the same operation name can exist for different types. As a shorthand, we therefore use $\texttt{op}_{\texttt{type}}$ to refer to the realization of the operator $\texttt{op}$ for the attribute type $\texttt{type}$. The operator "$\geq_{\mathbb{I}}$", for example, identifies the *greater than or equal* function for integer numbers in the prototype implementation of BYPASS.KOM. The set of available operations for a specific attribute type is referred to as $OP_{\texttt{type}}$.

Combining operators and attributes leads to the specification of filters. A filter f is defined as the tuple $(a, \texttt{value}, \texttt{op})$ of attribute, value, and operator. As in

[2]Some shorthand and utility methods are omitted for brevity. A complete documentation of the API available in the SIMONSTRATOR.KOM platform is provided online: www.simonstrator.com (Accessed March 8th, 2017).

the previous cases, entries within the tuple are addressed as $f.a$, $f.$`value`, and $f.$`op`. To probe an input value against the respective filter, the function referred to by $f.$`op` needs to be executed at the broker. This is denoted through the function `eval(`$f.$`op`, $f.$`value`, `value`), which evaluates the operation for the provided candidate `value` and returns a boolean indicating a match. Note, that the type of `value` must match $f.a.\texttt{type}$, and that $op_{f.a.\texttt{type}} \in OP_{f.a.\texttt{type}}$ must hold. Following our example, a filter for all events where the attribute "level" is greater than or equal to 12 is defined as ("level", 12, "≥").

Consequently, a subscription s is defined as a non-empty set of filters $s = \{f_1, \ldots, f_n\}$. To match an event e against a set of subscriptions S, Algorithm 1 is applied for each subscription $s \in S$. The algorithm first checks if the attribute used within each filter f_i is also contained within the event, as the event is otherwise not covered by the subscription. As the `eval` function is applied to all attribute-value pairs with the same attribute, a subscription can contain multiple filter entries for the same attribute. The algorithm returns true if, and only if, all filter entries are matched successfully.

Algorithm 1: Matching an event against a single subscription.

```
   Data: Event e and subscription s
   Result: Boolean indicating a match
1  for f_i ∈ s do                                   // per definition: s ≠ ∅
2  |  if f_i.a ∉ e.a then
3  |  |  return false;
4  |  for e_j ∈ e do
5  |  |  if e_j.a = f_i.a then
6  |  |  |  if not eval(f_i.op, f_i.value, e_j.value) then
7  |  |  |  |  return false;
8  return true;
```

Finally, the clients associated to the set of subscriptions $S' \subset S$ for which Algorithm 1 returned true are notified of the event e.

4.2.1.2 Subscribing to Events Around a Client's Location

As the subscription model in BYPASS.KOM does not contain any specific extensions for location-based subscriptions, we provide this functionality through a dedicated API method:

```
sub_handle subscribe(subscription, callback,
    location_request, radius)
```

In comparison to the `subscribe` method defined in [18], we added the underlined arguments `location_request` and `radius`. `Subscription` refers to a set of attribute-based filters, as defined in our subscription model. The `callback` is notified, whenever a matching event is delivered to the client. The method returns `sub_handle` as a unique identifier for the subscription that can later be used to unsubscribe.

As motivated, we require filter schemes to manage a client's location context themselves. Therefore, the application does not subscribe to a static location, but instead describes how continuous location updates should be handled by the filter scheme. Subscribing to a *static* location does not require any context management and can simply be achieved by adding a custom attribute type and the corresponding operator to BYPASS.KOM. Thus, it can be realized independent of the utilized filter scheme through the default subscribe method as defined in [18].

To enable an application to control the desired accuracy and frequency of updates, we demand it to pass a `location_request` containing this information. The location request is motivated by the design of the location sensor API of Google's Android mobile OS, which includes a corresponding `LocationRequest` object.[3] The `location_request` simply contains the desired frequency of location updates and the accuracy of the location information required by the application, usually specified as either *low*, *medium*, or *high accuracy*. For high accuracy or high frequencies, GPS or Wi-Fi fingerprinting is utilized by the Android OS. In the case of mobile augmented reality games, high location accuracy is a key requirement.

We assume that applications are interested in a circular AoI around the client's current location. The size of the AoI is defined by `radius`. Although being a simple integer value, this information is not encoded as an attribute within the subscription. As previously motivated, we separate contextual information associated with location-based publish/subscribe from our subscription model. Adding this client-specific attribute would hinder efficient merging of otherwise equal subscriptions within the (potentially channel- or attribute-based) broker network. Furthermore, many filter schemes utilize a custom representation of the AoI to deal with location uncertainties and low update frequencies [1]. This filter scheme-specific representation might deviate from the circular AoI used by the application through our API. It is up to the respective filter scheme how the radius as defined by the application is stored, communicated, and utilized within the filtering process, as later discussed in Sect. 4.2.2. We discuss the integration of such filter schemes into BYPASS.KOM in Sect. 4.2.4, demonstrating the applicability of our design.

4.2.1.3 Revoking a Location-Based Subscription

The signature of the `unsubscribe` method is left unchanged compared to [18]:

```
void unsubscribe(sub_handle)
```

[3]The full API of Android's `LocationRequest` is documented at https://developers.google.com/android/reference/com/google/android/gms/location/LocationRequest (Accessed March 8th, 2017).

However, if `sub_handle` refers to a location-based subscription, the client additionally checks whether it has to stop the associated request for location updates. This procedure depends on the client-side implementation of the utilized filter scheme and is therefore discussed in Sect. 4.2.2.

4.2.1.4 Publishing to a Location

In addition to supporting *traditional* events, BYPASS.KOM supports publishing an event that is relevant at the client's current location or at an arbitrary location. Publishing an event at the client's current location is performed by a dedicated API method:

```
void publish_local(event)
```

To publish an event to an arbitrary location, BYPASS.KOM further defines an extension of the `publish` method that accepts a geographical location.

```
void publish(event, location)
```

The distinction between publishing events to the client's current location and publishing to arbitrary locations is essential for efficient filtering, especially if channel-based filter schemes are utilized. This is further elaborated in Sect. 4.2.2. In both cases, the location is *not* part of the attributes defined in the event. Instead, its handling is solely defined by the respective filter scheme. Only events published through one of the aforementioned methods are subject to location-based filtering at the broker. All other events are filtered based on their static attributes.

4.2.2 Encapsulation of Filter Schemes

The filter schemes themselves are encapsulated as mechanisms at clients and brokers, as discussed in [21]. According to our design goals, this encapsulation has to be independent of the respective filter scheme to enable transitions. As illustrated in Fig. 4.2, each filter scheme consists of four building blocks: (i) a client component, (ii) a broker component, (iii) a protocol in between, and (iv) a storage for client context at the broker. We explicitly provide a distinction between the broker component of a filter scheme and its storage utilized for client context. As shown by Werner et al. [26] for routing protocols, a clear separation of protocol logic and the storage of associated state and context information aids in supporting state transfer and transformation during a transition. The influence of state transfer on the performance of BYPASS.KOM is evaluated in [21] and further discussed in Sect. 6.2.2.

In the following, we discuss the respective building blocks for two distinct classes of location-based filter schemes identified in our analysis of the state of the art. The schemes realized and evaluated within the prototype of BYPASS.KOM are described in detail in Sect. 4.2.4.

4.2.2.1 Integration of Context-Based Filter Schemes

Within our model, context-based filter schemes are realized as a combination of a *traditional* subscription as defined in Sect. 4.2.1 and context information associated to the respective subscriber or a subscription at the broker. For the case of location-based publish/subscribe, this context information consists of the last known location and the AoI of the subscriber. Incoming location-based events are matched against the set of location-based subscriptions by following Algorithm 1 for all subscriptions that are still valid candidates after additional matching rules for the associated contextual information have been applied. The order of these two steps depends on the implementation of the filter scheme, as the resulting efficiency is affected by the underlying data structures and the characteristics of the subscriptions [11, 28].

In addition to simply forwarding events and subscriptions to the broker, the client component is responsible for retrieving continuous updates of the mobile device's current location when issuing a location-based subscription. This is done through interaction with the mobile OS, as introduced in Sect. 2.1. The application specifies an expected accuracy and update frequency of location information as part of the location request provided through the API. As described previously, multiple parallel location requests are usually grouped by the mobile OS. Nevertheless, the client component has to decide when and how often location updates have to be sent to the broker to maintain correct operation of the filter scheme.

Within BYPASS.KOM a distance-based, a frequency-based, and a hybrid approach for context updates are realized. With the distance-based approach, a new location is sent to the broker as soon as the distance between the last reported location and this new location exceeds a predefined threshold. With the frequency-based approach, locations are reported periodically, regardless of the actual physical distance between two reported locations. The hybrid approach simply reports a location as soon as either one of the previous two approaches would send an update.[4]

In addition to location updates, context-based schemes might require the transmission of additional contextual data. This includes the size and shape of the AoI. For a circular AoI, clients need to report the radius for a new location-based subscription to the broker. As this information is static for a subscription, it only needs to be reported once. For advanced schemes such as STEs [1], each location update also involves transmission of a vector that denotes the speed and direction of the client's movement. While this information could also be extrapolated at the broker, direct access to sensors like compass and gyroscope lead to higher accuracy of client-reported information. Context-based subscription schemes rely on a unidirectional context update protocol. All updates are initiated by the client based on local knowledge obtained from sensor measurements.

Consequently, the broker component does not initiate any communication other than the usual forwarding of matched events to subscribed clients. Matching, as briefly discussed, involves scheme-specific operations on the stored contextual

[4] If a location is reported as consequence of the distance-based method, the frequency-based approach simply begins a new period to avoid duplicate location updates.

information. For the case of a circular AoI, matching simply involves calculating the distance between subscriber and target event location and comparing the result to the provided radius of the AoI. A geographical location l_i is specified as a pair $l_i = (\varphi_i, \lambda_i)$ of latitude and longitude, respectively. The distance $d(l_1, L_2)$ between two locations l_1 and l_2 is calculated at the broker using the haversine formula [25], with R denoting the Earth's radius and φ_x, λ_x given in radians:

$$d = 2R \arcsin\left(\sqrt{\sin^2\left(\frac{\varphi_2 - \varphi_1}{2}\right) + \cos(\varphi_1)\cos(\varphi_2)\sin^2\left(\frac{\lambda_2 - \lambda_1}{2}\right)}\right).$$

Within location-based applications, the radius of the AoI usually lies in the range of a few hundred meters. As discussed previously, we further assume that clients are already assigned to coarsely distributed brokers based on their location. Therefore, inaccuracies arising from calculations of very large distances using the haversine formula are not an issue within our work. As soon as schemes construct more complex shapes for their AoI, calculations are usually executed in the Cartesian coordinate space. A location (φ_i, λ_i) on the Earth's surface can be mapped to the corresponding (x_i, y_i) coordinate using the Mercator projection [17], with

$$x_i = R\lambda_i, \qquad y_i = R \ln\left[\tan\left(\frac{\pi}{4} + \frac{\varphi_i}{2}\right)\right].$$

Consequently, the distance d between two locations l_1 and l_2 is now simply calculated as the euclidean distance

$$d = \sqrt{(x_2 - x_1)^2 + (y_2 - y_1)^2}.$$

The Cartesian coordinate space is, for example, utilized within our prototype when calculating the conical extensions of the AoI for the STE scheme [1].

To perform these calculations, all context-based filter schemes need to store the last known location of subscribers. Additionally, the AoI needs to be stored for each location-based subscription. If client mobility is to be extrapolated by the broker, further information such as timestamps of the last update might need to be maintained. As this additional information is decoupled from the subscription itself due to the design of our subscription model, we do not pose any restrictions on the context information associated to a subscription and/or a subscriber. Following this blueprint, we integrate state of the art approaches for context-based filtering into our prototype of BYPASS.KOM, as later presented in Sect. 4.2.4.

4.2.2.2 Mapping Locations to Channel-Based Filter Schemes

Utilizing a channel-based filter scheme provides the benefit of increased efficiency at the broker and enables the utilization of lower-layer multicast technologies when distributing events to a larger set of subscribers, as already discussed in Sect. 3.1.3.

To integrate channel-based schemes into BYPASS.KOM, we follow the same methodology as for context-based filter schemes by dividing the filter schemes into client and broker components that communicate via a scheme-specific context update protocol. All channel-based schemes for location-based publish/subscribe strive to solve two fundamental problems: (i) an efficient initialization of channels, and (ii) a lightweight assignment of clients to channels.

The initialization of channels is done by the broker component of the respective filter scheme. As discussed in Sect. 3.1.3, channels can simply correspond to a geometric overlay such as a grid or hexagonal cells with predefined (although not necessarily static) boundaries. However, more complex schemes can benefit from available map data in assigning clients to channels. Thereby, channels can be initialized based on specific real-world places of interest, e. g., for larger events. Channels do not need to be static. Instead, the broker might decide to alter the organization of channels depending on load or any other internal or external factor.

Given the complete set of channels C spanning the geographical area of responsibility of a given broker, the next task is to assign clients to one or multiple channels depending on their current location. Obviously, the assignment cannot be static in the dynamic scenario of mobile location-based applications. Once a client leaves the region of coverage of a channel, the assignment process has to be repeated for that client. We introduce the *assignment function* `assign`: $(C, l, r) \rightarrow (C', c_p)$ with $c_p \in C'$ and $C' \subset C$. The function returns the set of channels C' out of all channels C that a client needs to be subscribed to when it is interested in events that are published to location l with a radius of r. Additionally, a channel c_p that is to be used for all outgoing local events is returned. This function is realized within the broker component, as it requires full knowledge of all available channels. Depending on the filter scheme, additional (potentially application-specific) input other than the client's current location can be utilized to perform the assignment.

Knowledge about the complete set of channels C and their geographical boundaries is usually not available at the client components. Instead, a client initially retrieves a local view consisting of (at least) the assigned channels C' from the broker. This local view can either be requested explicitly by the client, for example by sending the current location to the broker, or it can be pushed to currently subscribed clients by the broker, for example as consequence of a re-initialization of the available channels. Filter schemes aim to minimize the communication overhead associated with acquiring the assigned channels. Therefore, clients do not report their location periodically, but instead perform local tests based on C' and any additional local knowledge available, on whether they need to contact the broker to update their assignment. This is discussed in more detail in Sect. 4.2.4.

As discussed, the API for location-based publish/subscribe defined in BYPASS.KOM distinguishes between events targeted at the client's current location and events for arbitrary locations. This distinction is essential for channel-based filter schemes. If an event is targeted at the client's current location, it is published to the channel c_p. Thereby, filtering at the broker component does not involve any additional calculations related to the location information.

However, for arbitrary target locations, the channels that are to be notified are usually not known to the client. Determining the target channels for a location would require local knowledge of the broker's assignment function, which is usually not the case. Therefore, the event together with its target location have to be sent to the broker, where the respective channels are then determined by applying the assignment function to the target location. This process can introduce significant computational overhead at the broker, thereby counteracting the benefits usually associated with channel-based schemes. This motivates the execution of transitions depending on the workload characteristics. As soon as the fraction of events published to arbitrary locations exceeds a threshold, a context-based filter scheme could be utilized instead.

4.2.2.3 Multiplexing Filter Scheme Messages

To fully encapsulate filter schemes, we also need to consider the utilization of network resources. This is especially important during a transition, where two filter schemes can be active at a client at the same time, as discussed in the following section. Therefore, filter scheme specific messages used by the context update protocol are multiplexed. The *network proxy* component at brokers and clients adds a flag to outgoing messages, indicating the currently active filter scheme of the sender. Relying on this flag, the receiving entity's network proxy component is able to determine the corresponding instance that has to be notified of the incoming message.

The events and subscriptions themselves are not scheme specific. This is due to our separation between subscription model and associated contextual information. Consequently, they are sent via a common protocol regardless of the filter scheme. This mechanism ensures that events and subscriptions issued during a transition from one scheme to another can still be processed by the broker. Nevertheless, the network proxy adds a flag to identify the currently active filter scheme. This information can be utilized to detect failed client transitions and is required if the broker operates multiple schemes in parallel to cater for different groups of clients. The latter is discussed in the context of coexisting mechanisms in Sect. 4.4.

As most filter schemes associate context to a subscription once the subscribe method is invoked, BYPASS.KOM supports piggybacking of messages. If a scheme's client component wants to associate contextual information to an outgoing subscription or event, the respective context update message is simply appended to the basic message. This reduces the overhead caused by lower-layer message headers, which is especially important if context update messages carry only a few bytes payload.

4.2.3 Executing Transitions Between Filter Schemes

In addition to the encapsulation of individual schemes into broker- and client components and the corresponding context update protocol, we need to control each component's lifecycle to execute transitions. By relying on generic lifecycle methods

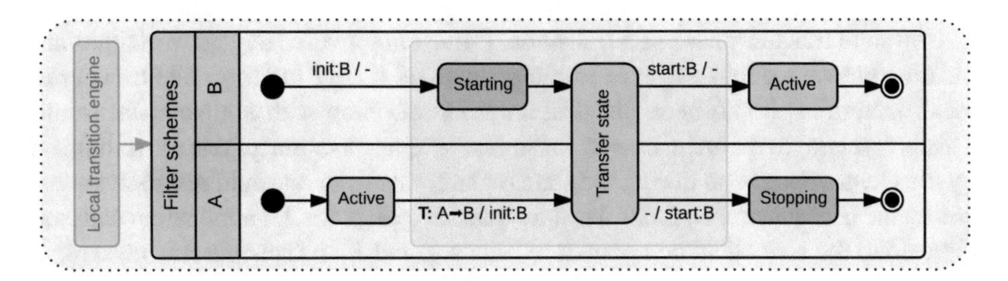

Fig. 4.3 Execution of the filter scheme transition $T : A \rightarrow B$

as later described in Sect. 5.2.1, we can start and stop specific mechanism instances during a transition without utilizing any mechanism-specific functionality. In the following, we focus on the local execution of a transition once a trigger is received, as illustrated in Fig. 4.3. The local transition engine is a generic component that is later discussed in Chap. 5. For now, we simply assume that a transition is to be executed and that clients are notified of this decision.

Here, we have to distinguish between two fundamentally different scenarios resulting from the target scope of a transition. A transition can either affect all clients associated to a broker (which we refer to as a *total* transition), or just a subset of clients (a *partial* transition). In [22], we focused on total transitions, arguing that a broker is responsible for a relatively small and locally confined set of clients. The resulting consequences on the execution of a transition are discussed in the following section. Afterwards, we consider the case where transitions affect only a subset of clients assigned to a broker.

4.2.3.1 Total Transitions on All Clients

In the following, we refer to the total transition from one filter scheme A to another filter scheme B as $T : A \rightarrow B$. The process of applying T to the client component is discussed hereafter.

First, the client component of the new filter scheme B is initialized. At this stage, all interactions with the client's filter scheme are still directed to scheme A. After B is successfully initialized, relevant state information from A is transferred to B. This step might involve a transformation of the state to fit B. This process is referred to as the *state transfer* phase of T.

Relevant state of a filter scheme's client component includes the client's subscriptions and corresponding callbacks. Without transferring this information from one scheme to the other, the client application would have to resubscribe after each transition. Consequently, transitions would no longer be transparent to the application and additional overhead is caused, as shown in Sect. 6.2. As we rely on a common subscription model for all filter schemes, transferring subscriptions and their application callbacks does not require transformation.

The state transfer phase of a transition T must not involve any network communication between client and broker components of B. By limiting state transfer to local operations, it can be realized in an atomic fashion with a guaranteed result. This is essential to maintain consistent state: if events or subscriptions are initialized by the client application during the state transfer phase, they would not be available within the new scheme B. After the state transfer took place, all application calls are directed to the new client component of filter scheme B and scheme A is discarded. At this stage, the transition finished successfully.

Still, some schemes require additional bootstrapping that involves contacting the broker component. One example are channel-based schemes that require initial assignment of channels to the client before being able to process new events efficiently, as discussed in Sect. 4.2.2. This is referred to as the *bootstrap phase* of a filter scheme. As the bootstrap phase takes place after the transition has finished, it is no longer formally a part of the transition. More specifically, the bootstrap phase of a filter scheme is usually realized in the same way as an initialization of a scheme.

Executing a total transition at the broker does not differ from the previously described process for client component transitions. As the transition affects all clients, we do not need to maintain parallel operation of multiple different filter schemes. Instead, the new scheme B is initialized, state is transferred locally, and scheme A is stopped afterwards. State transfer involves moving all stored subscriptions from scheme A to scheme B.

Additionally, context such as the last known location of a subscriber can be transferred during the state transfer phase. Whether such context information is available or usable during a transition depends on the respective filter schemes. Context information that is required by B but not available within A has to be gathered during the bootstrap phase of the filter scheme. This process can be initiated by the client component or the broker component, as previously discussed.

As for the client component, the state transfer phase of the transition at the broker is atomic and must not contain any network communication. However, given that some schemes might require contextual information to ensure correct location-based filtering of events, we propose three strategies to handle events that arrive during the bootstrap phase:

Filter without context. If the overall latency of the event delivery is to be minimized, we simply treat location-based subscriptions as normal subscriptions until the context information has arrived. Thereby, the event is forwarded to all matching subscribers regardless of their actual location, requiring additional filtering at the client. This leads to increased overhead due to unnecessary transmissions.

Wait for context. Here, incoming events are stored in a buffer while the bootstrap phase has not yet finished for a subscriber. Once the context information arrives, the events are filtered and matching events are delivered to the client. This method ensures that only relevant events are delivered to clients. However, the overall latency for event delivery increases.

Ignore. This strategy ignores the respective subscription until required client context has arrived. This is only applicable if the application can tolerate potential loss of events during a transition. It does not infer any overhead or additional complexity at the broker.

In the prototype of BYPASS.KOM, we ignore subscriptions with incomplete contextual information. In Sect. 6.2.2, we show that by applying our state transfer mechanisms, we can still ensure reliable operation during most transitions. However, if reliable delivery of events is to be guaranteed, one of the other strategies is to be selected based on the tolerable delivery delay.

4.2.3.2 Partial Transitions on a Subset of Clients

If transitions between filter schemes are to be executed only for a subset of the assigned clients, this affects the execution of the transition at the broker component. Instead of applying the total transition T, the partial transition $T^{\parallel}$ is executed at the broker as discussed in the following. Clients simply execute the total transition T, as described in the previous section.

At the broker, two or more filter schemes need to be running in parallel, if subsets of clients use different schemes as consequence of a partial transition. Incoming events need to be processed by all active schemes to guarantee delivery to all interested clients. Therefore, the network proxy introduced in Sect. 4.2.2 duplicates the respective incoming message and forwards it to all active filter schemes. Subscriptions issued by clients are only processed by the scheme that is currently active at the client. The flag attached to outgoing messages at the client is used by the network proxy to determine the correct filter scheme.

How a partial transition $T^{\parallel}(u)\forall_u \in U'$ between scheme A and B is executed for a set of clients U' depends on whether scheme B is already running at the broker. Assuming that this is not the case, the broker component of scheme B is initialized, and the state transfer phase is started. However, in contrast to a total transition, only state that is associated to clients $u \in U'$ that are affected by the transition $T^{\parallel}$ is transferred. As is the case for the total transition T, state includes all subscriptions of the respective clients and, optionally, contextual information managed by the filter scheme. As soon as the state transfer is finished, scheme B is actively utilized by the network proxy as described previously and starts processing incoming events.

If scheme B is already active when a transition is to be executed, the initialization phase is skipped. In both cases, bootstrapping for the newly assigned clients can occur, as is the case for a total transition. If scheme A is no longer utilized as a consequence of $T^{\parallel} : A \rightarrow B$, its broker component is discarded. The realization of this process is described in more detail in Sect. 5.2.

4.2.4 Integration of Existing Mechanisms

The modular framework BYPASS.KOM supports the integration of existing mechanisms to research on transitions in between these mechanisms. In this section, we discuss how representative mechanisms from the state of the art in location-based filtering are integrated into our framework. In particular, we focus on the separation into client and broker component and the context update protocol. We address how the respective filter schemes utilize the common subscription model combined with custom context handling to realize their functionality in BYPASS.KOM. Further, we discuss alternatives for the organization of subscriptions and the utilization of contextual information in distinct filter schemes. We provide a brief discussion on how the respective alternatives affect the complexity of the filter process at the broker.

4.2.4.1 Parametric Subscriptions

As perhaps the most intuitive filter scheme for location-based publish/subscribe, we integrate parametric subscriptions [12] into BYPASS.KOM, hereafter referred to as RADIAL. Within RADIAL, filters $\{f_1, \ldots, f_n\}$ of a location-based subscription are augmented with the parameterized filter expression

$$f_p = \left(d\left(\$(l_u), l_e\right),\ \$(\text{radius}),\ \text{“}\leq\text{”}\right) .$$

We refer to the function $\$(\cdot)$ as *context resolver*. When the filter is to be evaluated, the context resolver inserts the current value for the context variable. In case of $\$(l_u)$, this is the last known location of the client as stored on the broker. The location l_e is contained within the event e that is to be matched against the subscription and does not need to be resolved by the broker. The radius of the subscription is stored as associated state information for a client's subscription and is therefore also resolved using $(radius). The function $d(\cdot)$ returns the distance between two locations as defined in Sect. 4.2.2.

To enable location-based filtering, the broker component of RADIAL has to maintain (i) the last reported location of a subscriber, l_u, and (ii) the radius associated to a subscription, radius. The last known location is valid for all location-based subscriptions of the respective client, while the radius needs to be associated to the client *and* the respective subscription.

When subscribing to a location through our API, the requested radius is reported to the broker component as part of the context update protocol. Additionally, the client component requests the client's current location and following updates of this information according to the location request using the API provided by the mobile OS. The respective context updates are triggered by the client as described in Sect. 4.2.2. Consequently, the RADIAL scheme's context update protocol consists of a single message type, `CtxUpdateMsg`. The `CtxUpdateMsg` carries one or more key–value pairs for context variables that are to be updated at the broker. Given that

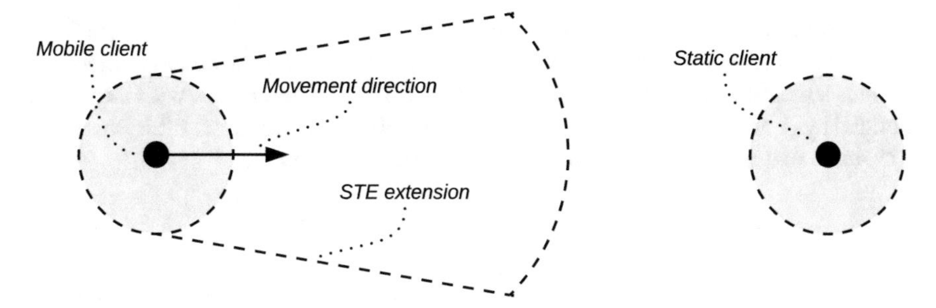

Fig. 4.4 Area defined by a STE

the radius associated to a subscription does not change[5] over time, it is only reported once upon creation of the new subscription.

It is important to note that parametric subscriptions as defined by Jayaram et al. in [12] and implemented in BYPASS.KOM are not limited to location-based filtering. Any contextual information can be included in a parameterized filter, given that the respective context can be resolved at the broker.

4.2.4.2 Space-Time Envelopes

As proposed by Brimicombe et al. in [1], STEs extend the concept of matching against a circular AoI by taking the client's current movement speed and direction into account. Thereby, as discussed previously, clients need to issue less frequent location updates and the filter scheme is more robust against higher movement speeds. We refer to our realization of the scheme within BYPASS.KOM simply as STE. To calculate the area covered by an envelope in STE, the broker needs to know (i) a subscriber's current location l_u, (ii) the subscriber's current movement vector $\mathbf{s}_u$, and (iii) the radius associated to the subscription.

The area is then constructed as illustrated in Fig. 4.4: the circle defined by the client's current location l_u and the radius of the subscription is extended with a cone in the direction of $\mathbf{s}_u$. The cone's initial width is defined by the radius, widening with an angle of 15°. Its length is given as $\alpha \cdot 60 \cdot |\mathbf{s}_u|$, with $|\mathbf{s}_u|$ being the client's movement speed in meters per second. The factor α can be used to adapt the scheme to different movement speeds and location update intervals. Its default value is $\alpha = 1$, corresponding to a cone covering roughly one minute of unaltered client mobility.

To reduce the computational complexity at the broker, the respective area is calculated once after an update of any of the above-mentioned parameters and then stored in the context storage. As the area depends on the radius associated to a subscription, the STE needs to be stored as subscription-dependent context. Thus, we refer to the STE of client u and subscription s as $STE_{u,s}$.

[5] As defined by our API, applications specify a fixed radius when issuing a location-based subscription. To alter the respective AoI, the application simply resubscribes with an altered radius.

Incoming events are matched against location-based subscriptions by checking whether their location l_e lies within the area $STE_{u,s}$. To this end, the broker component defines a method `in_area`(A, l) that returns 1 if the location l lies within the area defined by A, or 0 otherwise. Consequently, the following additional filter is used by STE when matching events against location-based subscription:

$$f_{STE} = \left(\text{in\_area}\left(\$(STE_{u,s}), l_e\right),\ 1,\ \text{“=”}\right) .$$

Note, that the current area of the STE for client u with subscription s is resolved as a context variable via $\$(\cdot)$.

In both schemes, RADIAL and STE, custom functions are utilized, thus increasing the complexity of the matching process at the broker. Especially for STE, determining whether a given location lies within the area covered by an STE is a rather computationally intense task. In addition to the computational complexity associated with context-based subscriptions, context variables have to be updated frequently to ensure accurate filtering at the broker. To lower the complexity of the filtering process and the frequency of context update messages, the following channel-based filter schemes are integrated into BYPASS.KOM.

4.2.4.3 Grid-Based Filter Schemes

Channel-based schemes aim to reduce the communication overhead caused by frequent context update messages. At the same time, they utilize pure attribute-based filtering at the broker, thereby reducing the complexity of the filtering process. Based on the general discussion of channel-based filter schemes in Sect. 4.2.2, we discuss three distinct representatives integrated into BYPASS.KOM. First, we discuss a basic grid with single-channel assignment, called GRID, followed by the multi-channel assignment scheme EGRID. Motivated by the characteristics of location-based mobile applications, we then propose a channel-based scheme called ATTRACT that utilizes application-specific knowledge to optimize its channel assignment.

In GRID and EGRID, channels are organized as a grid of equally sized rectangles covering the area that is managed by the broker, hereafter referred to as W. For simplicity, we assume that a client's location l_u and the bounds of W are transformed into the Cartesian coordinate space using a suitable projection as discussed before. A grid is constructed based on the MBR around the area W. This MBR is divided into equally sized rectangles, with the number of divisions for each axis being configured with the so-called *grid factor*. A grid factor of 2, for example, leads to a 2×2 grid and, consequently, 4 channels.

As defined in Sect. 4.2.2, the assignment function returns the respective set of channels a client needs to be subscribed to. Within GRID, a client is assigned only to the channel covering the client's current location, regardless of the AoI of any subscriptions. Within EGRID, the client is instead assigned to all channels within the MBR required to cover the AoI, as illustrated in Fig. 4.5. Therefore, the broker additionally needs to store the size of the AoI of a subscription to provide the context

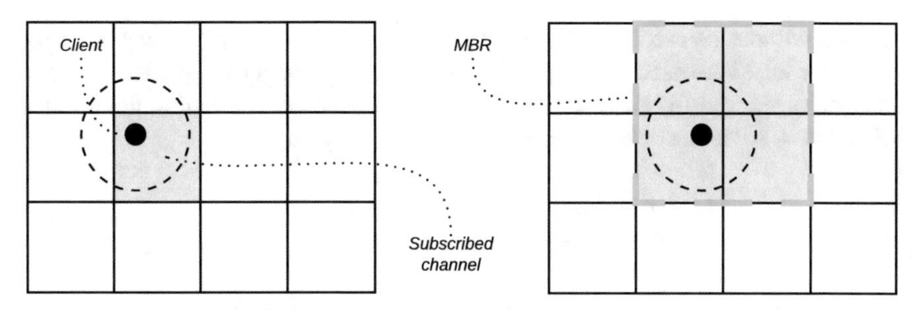

Fig. 4.5 Channel creation and assignment in GRID and EGRID

variable \$(radius) required by EGRID. While this information is not actually required for the operation of GRID, it is nevertheless maintained as context information as it contributes to the seamlessness of transitions as evaluated in Sect. 6.2.2.

A client needs to send a `RequestAssignmentMsg` to the broker to be assigned to a channel. The message simply contains the client's current location l_u, much like the aforementioned context update messages. Based on the reported location, the broker assigns the channel c_p to be used for outgoing local events and notifies the client of the respective channel name and the geographical boundaries of the MBR by sending an `AssignmentMsg`. Based on the MBR, the client can detect locally whether a new assignment needs to be requested as consequence of mobility. An `AssignmentMsg` is also issued by the broker whenever the assignment function itself is updated, for example due to an updated grid factor.

The full set of channels C' a client is subscribed to is stored as contextual information at the broker. As a client u can have multiple active location-based subscriptions with different AoIs, this context variable also depends on the respective subscription s. It can, therefore, be accessed with $\$(C'_{u,s})$. The client component adds c_p to outgoing *local* events by adding a new attribute

$$a_c = \left(\text{“channel”},\ c_p\right).$$

This attribute is used by the broker during the matching process. The broker component applies the filter

$$f_c = \left(\text{“channel”},\ \$(C'_{u,s}),\ \text{“}\in\text{”}\right)$$

to all location-based subscriptions during matching. The set operation "∈" checks whether $c_p \in C'_{u,s}$ by comparing the respective channel names. It does *not* involve any geometric calculations. As matching does not involve a custom function in this case but instead relies on a simple hash-based comparison of channel identifiers, it is considered to be more efficient in terms of computational overhead than the previously discussed context-based schemes.

The aforementioned realization relies on the context resolver $\$(\cdot)$ to retrieve the set of channels for a client and its subscription. The same functionality can also be realized without calls to $(·) during filtering by creating $k = |C'_{u,s}|$ copies of the respective location-based subscription, with each copy s_i being assigned to one fixed channel $c_i \in C'_{u,s}$. In this case, each copy s_i is created by extending the original subscription s with a context-independent filter f_{c_i}, leading to:

$$s_i = \left\{ f_{c_i} \right\} \cup s \quad \forall i \in \{1, \ldots, k\} \quad \text{with}$$
$$f_{c_i} = (\text{"channel"},\ c_i,\ \text{"="})\ .$$

With this, the context resolver is not utilized when matching incoming events against subscriptions. However, the set of subscriptions has to be altered according to the updated assignments whenever a client issues a new `RequestAssignmentMsg`, now involving the context resolver. In our application scenario, the rate of events and consequent matching operations is much higher than the rate of `Request AssignmentMsg`, especially if channels cover a sufficiently large area. Additionally, while the total number of subscriptions increases for the context-independent subscriptions, they can be organized in data structures that enable lookup of all subscriptions for a given target channel with an average $\mathcal{O}(1)$ time complexity for non-overlapping channels [14]. In case of overlapping or nested channels, more elaborate index structures as proposed by Zhou et al. in [29] can still be applied. Compared to RADIAL and STE, this leads to location-based filtering with lower computational complexity, given that all schemes still perform attribute-based matching as outlined in Sect. 4.2.1.

A client cannot utilize the channel c_p as described above, if outgoing events are published to *arbitrary* locations. Instead, the event together with its target location l_e needs to be sent to the broker, which then determines the correct target channel and adds the corresponding attribute a_c. Afterwards, matching is performed as discussed above for local events.

The broker does not need to store any additional client context for channel-based schemes. However, to realize seamless transitions between filter schemes, the last known location of each client is maintained within BYPASS.KOM. This information is also helpful during self-transitions used to adapt the size of the grid and, consequently, the current set of channels. Self-transitions are later discussed in Sect. 4.2.5 and their utilization for adapting the grid sizes in channel-based schemes is evaluated in Sect. 6.2.3. In addition to grids, we study an application-specific approach to the assignment of channel areas in the following section.

4.2.4.4 Utilizing Application Knowledge with Attraction-Based Channels

The GRID and EGRID filter schemes are application agnostic. With ATTRACT, we extend the previously discussed channel-based filter schemes by specifying an application-specific assignment function. In ATTRACT, channels are defined based on application-specific attraction points to better capture the real-world behavior of

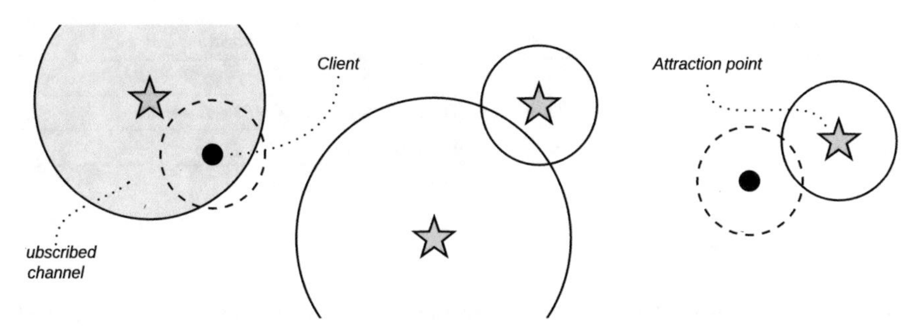

Fig. 4.6 Channel assignment based on application-specific attraction points

clients in mobile social applications. Attraction points are modeled as a circular area, with their center locations and radii being given, as later detailed in Sect. 6.1.1.

Within ATTRACT, we simply assign a channel to each attraction point rather than relying on a fixed grid as discussed in the previous section. The area covered by the channel is defined by the radius of the respective attraction point, as illustrated in Fig. 4.6. In contrast to the previously discussed grid-based filter schemes, channels in ATTRACT can overlap. Additionally, depending on the locations and radii of attraction points, some locations on the map are not covered by a channel at all. In case a location is covered by multiple channels, the client subscribes to all channels. Consequently, the subscription at the broker is copied for each channel as discussed for the EGRID filter scheme in the previous section.

We propose two distinct methods to deal with locations that are not covered by a channel at all. A simple solution is to define a distinct channel used in that case, hereafter referred to as c_{other}. Obviously, this greatly reduces the precision of event brokering for clients outside the area covered by attraction points. Additionally, the area covered by c_{other} cannot be described as a simple MBR or circle. Consequently, either the client or the broker has to check periodically whether the client's current location is still not covered by another channel. The former requires local knowledge about the areas covered by all channels, while the latter requires periodical updates of a client's location. Given that we deploy channel-based schemes to reduce the computational overhead at the broker, we use the client-side approach in ATTRACT. However, periodic location updates can be enabled for ATTRACT with the self-transition $T^{\bullet} : f(\text{ATTRACT}) = \text{enable\_updates} \leftarrow \text{true}$, as utilized in conjunction with gateway-based event dissemination presented in Sect. 4.4. The concept of a self-transition is detailed in the following section.

Another way of dealing with locations that are not covered by a channel is the utilization of parametric subscriptions in this case. Therefore, clients that are not covered by a channel report their current location periodically and the broker filters events according to the procedure defined for RADIAL. Which strategy to use depends on the scenario at hand, especially with respect to the density and expected radii of attraction points. Within this work, we use the combination of ATTRACT with the parametric scheme RADIAL, referred to as MULTI, to compare the performance

Table 4.1 Transition types and notations

Type	Notation
Total transition (Sect. 4.2.3)	$T : A \rightarrow B$
Partial transition (Sect. 4.2.3)	$T^{\parallel}(u) : A \rightarrow B$
Self-transition (Sect. 4.2.5)	$T^{\bullet} : f(A)$

of a single adaptive filter scheme against that of a transition-enabled system. The transition-enabled system realizes the same behavior by executing the respective partial transitions between ATTRACT and RADIAL whenever a client enters or leaves the area covered by an attraction point. This is discussed and evaluated in Sect. 6.2.4.

4.2.5 Mechanism Reconfiguration Through Self-transitions

Most of the aforementioned filter schemes can be adapted to slight changes in the environmental conditions by altering configuration parameters. Within STE, for example, the extension of the envelope can be controlled with the parameter α, as discussed before. For GRID and EGRID, the grid size could be altered. Within BYPASS.KOM, we want to leverage the potential of such self-adaptation mechanisms in addition to the execution of mechanism transitions.

Parameter changes are modeled as *self-transitions*. A self-transition $T^{\bullet} : f(A)$ is used to alter the state of a mechanism A by applying a custom function $f(\cdot)$ to the currently active mechanism instance. Within BYPASS.KOM, we realize the adaptation of filter schemes as self-transition at the broker component of the respective filter scheme. During a self-transition, the respective configuration parameter of the running mechanism instance is updated. The generic realization of self-transitions within the SIMONSTRATOR.KOM platform is discussed in Chap. 5.

In the space-time envelope filter scheme STE, the current value of α is only required at the broker component. Clients just report their location and movement vector and all calculations of the resulting envelope are done by the broker. However, for the grid-based schemes GRID and EGRID, updating the grid size leads to an updated channel assignment. Therefore, as a consequence of self-transitions on the broker, the respective filter schemes utilize their custom context update protocol to inform clients of the newly assigned channels.

It is important to note that the self-transition concept is designed such that it utilizes a scheme's context update protocol. This is due to the observation that most existing mechanisms are already designed with some key properties such as self-healing and self-adaptation in mind. As is the case for the general concept of a transition, we want to utilize such existing properties whenever possible.

As discussed, the encapsulation of filter schemes as mechanisms in BYPASS.KOM and the utilization of a common subscription model enables the execution of transitions between different such schemes. To ensure that the performance of the publish/subscribe system is not degraded while transitions are being executed, we proposed mechanisms for state transfer. Furthermore, we explicitly addressed the potential of including application-specific knowledge and limiting transitions to a specific group of clients by means of partial transitions. We described the integration of distinct classes of filter schemes into BYPASS.KOM, enabling us to evaluate the impact of transitions under unified conditions, as later discussed in Chap. 6. We introduced self-transitions as a generic concept to reconfigure and adapt a mechanism, allowing more fine-grained control over a mechanism's behavior. Table 4.1 provides an overview over the transition types introduced in the previous sections. They are further generalized in the SIMONSTRATOR.KOM platform, as later discussed in Chap. 5.

Up to now, we addressed the challenge of adaptivity in location-based filtering. In the following section, we extend the discussion to event dissemination mechanisms and their transitions, addressing adaptive locality-aware event dissemination.

4.3 Transitions Between Event Dissemination Mechanisms

In addition to location-based filtering, we aim to achieve locality-aware event dissemination to exploit the locality in user interaction within the communication system. Figure 4.7 shows how events are processed by the locality-aware dissemination mechanisms of BYPASS.KOM at the client and the broker. The respective functionality depends on the rules applied at the client for instances of ⑦, as detailed later in this section. To address the challenge of efficiently utilizing a mobile device's capabilities in terms of direct local communication, we study three distinct approaches to locality-aware event dissemination within BYPASS.KOM: (i) direct local dissemination initiated by publishers (Sect. 4.3.1), (ii) hybrid local and infrastructure-based dissemination (Sect. 4.3.2), and (iii) locality-aware hybrid event dissemination using gateways (Sect. 4.3.4).

Locality-aware dissemination does not need to involve a central broker at all but can rely purely on a MANET to distribute and filter events instead. However, pure local dissemination is not feasible if global state is to be maintained at a central entity and for a globally distributed application. It can still be utilized to disseminate events that increase the perceived service quality without affecting global state, as presented in [19], or for applications that operate only within a spatially limited scope [20].

Consequently, for our scenario of location-based mobile social applications, we focus on infrastructure-based and hybrid event dissemination mechanisms, as discussed in Sect. 4.3.2. Given the research challenge of efficient utilization of the cellular connection, we further introduce mechanisms to reduce the load on the cellular infrastructure during the dissemination of events. To this end, we proposed a stateless approach to gateway selection in [23], detailed in Sect. 4.3.4. Our approach utilizes different selection algorithms and switches between gateways at no additional cost.

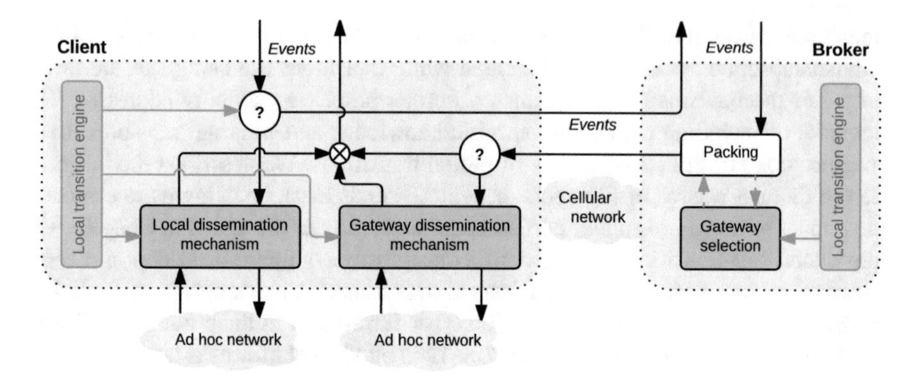

Fig. 4.7 Event dissemination components in BYPASS.KOM

This results in a highly adaptive and efficient locality-aware event dissemination in dynamic scenarios.

4.3.1 Local Ad Hoc Event Dissemination

If events are known to be relevant within proximity of their producer, local ad hoc dissemination mechanisms can be utilized to forward them to interested subscribers in proximity. Thereby, locality-aware dissemination is achieved, as events do not need to be processed at a centralized broker entity. Instead, clients themselves filter incoming events against their own subscriptions. Within our application scenario, subscriptions strongly depend on the subscriber's location and are, thus, subject to frequent updates. Therefore, we focus on the efficient dissemination of events while keeping subscriptions locally at each client.

In scenarios with more static subscriptions or less dynamic client mobility, one could also disseminate subscriptions instead. In this case, clients creating an event can already determine the set of interested subscribers and limit event delivery accordingly, as proposed in [10]. A combination of both approaches would involve rendezvous-based matching of subscriptions, as utilized in [5, 27]. We studied the respective characteristics in [19, 20], confirming that pure dissemination of events outperforms structured or rendezvous-based approaches in a dynamic scenario.

Although pure ad hoc dissemination of events cannot scale to a global scenario, we discuss dissemination mechanisms in BYPASS.KOM as the foundation for our hybrid approaches detailed in Sect. 4.3.2. Dissemination mechanisms offer a simple API to send an event to nearby clients:

```
void disseminate_locally(event, targets)
```

The second argument, `targets`, can be used to provide a set of target subscribers of the current event to the dissemination mechanism. The target subscribers are not known to a client when publishing an event, and they are not required for broadcast-based dissemination mechanisms. However, when acting as a gateway as discussed in Sect. 4.3.4, the cloud-based broker can optionally include the respective information to utilize unicast-based dissemination.

To be informed of incoming events, the application (or other components within BYPASS.KOM) can register as a listener:

```
void add_listener(listener)
```

Each time an event is received by the respective dissemination mechanism, the corresponding method is invoked on all listeners:

```
listener.event_arrived(event)
```

Additional methods are provided to remove listeners, if they are no longer utilized. To enable parallel operation of multiple instances of transition-enabled dissemination mechanisms, the port used for listening for incoming messages can be configured upon initialization. This is later utilized to distinguish between dissemination mechanisms used for direct dissemination of events and those used by gateways, as discussed in Sect. 4.4.

In the following, we briefly discuss the ad hoc event dissemination mechanisms that are available within BYPASS.KOM. As discussed in our analysis of the state of the art, we do not aim at developing new dissemination mechanisms. Instead, existing mechanisms are integrated into BYPASS.KOM to study their potential transitions.

4.3.1.1 One-Hop Dissemination via Unicast and Broadcast

As a basic primitive for local dissemination, unicast and broadcast via ad hoc network connections are supported within BYPASS.KOM. These mechanisms do not require any coordination between mobile clients prior to data transmission. However, assuming that the MANET does not utilize a network routing protocol, messages can only reach recipients within direct transmission range of the sender. While unicast messages are only processed by the intended receiver, broadcast messages are processed by all clients within transmission range.

Given the dynamic nature of the underlying network, we utilize UDP as basic primitive for all message transmissions. While broadcasts cannot be realized via TCP, direct unicasts in a dynamic environment would also suffer from TCP's bandwidth-probing approach and frequent disconnects due to client mobility, as shown by Fu et al. in [6]. Consequently, all dissemination protocols integrated within BYPASS.KOM utilize UDP broadcasts and unicasts as their basic messaging primitives. This way, the mechanisms can be utilized on all IP-based network stacks for direct ad hoc communication, including Wi-Fi and Bluetooth. One-hop communication can be utilized by selecting suitable gateways that inject information into their neighborhood. This is discussed in Sect. 4.3.4.

4.3.1.2 Multi-hop Event Dissemination

To disseminate information in larger ad hoc networks, multi-hop communication is necessary. Based on our survey of ad hoc dissemination mechanisms in Sect. 3.1.2, we integrate four multi-hop dissemination protocols as transition-enabled mechanisms into BYPASS.KOM. These mechanisms vary in their complexity in terms of state management and attempted re-transmissions of messages. In general, we aim to cover the full spectrum of low-complexity, low overhead, and potentially lower reliability up to high complexity, high overhead, and higher reliability.

Flooding. With FLOODING, each event is re-broadcast once on reception by each client. To detect and discard duplicates, each message is tagged with a UUID on creation. FLOODING causes significant load on the MANET, especially in densely populated areas. However, the aggressive dissemination strategy leads to high reliability as long as the network is not overloaded, as evaluated in Sect. A.3. In addition, the mechanism is robust against mobility, as no topology or state information about a client's neighborhood needs to be maintained.

Probabilistic Gossip. To reduce the load on the network while at the same time maintain the robustness properties of FLOODING, messages in GOSSIP are forwarded only with a given probability p. For $p = 1$, the mechanism behaves exactly like FLOODING, given that in GOSSIP we also utilize UUID to discard duplicate messages. There exist countless variations of gossip-based mechanisms in the literature, as discussed in Sect. 3.1.2. We chose a simple probabilistic approach given that we want to maintain low delivery delays. Still, by executing self-transitions as defined in Sect. 4.2.5, we can adapt the respective forwarding probability during runtime.

Hypergossiping. Proposed by Khelil et al. in [13], HYPERG is a self-adaptive dissemination mechanism based on GOSSIP. In contrast to GOSSIP, HYPERG adapts the probability of forwarding a message based on the current client density within proximity. To this end, a client within HYPERG periodically broadcasts a `Hello` message, allowing other clients to infer their approximate neighborhood based on the received `Hello` messages. Additionally, HYPERG utilizes a random hesitation time before rebroadcasting a message. If a configured amount of copies of the respective message are received during that hesitation time, the message is simply discarded.

Plan-B. Holzer et al. propose PLAN-B in [9]. In contrast to the aforementioned dissemination mechanisms, PLAN-B relies on location information to adapt the forwarding behavior of clients. When broadcasting a message, the sender piggybacks its physical location to the message. Receivers calculate the distance between the sender and their own location, based on which they determine a hesitation time. The hesitation time is chosen such that clients that are located farther away from the sender send the message first. Thereby, PLAN-B intends to cover a large area with only a minimum number of forwarders. In addition, clients in PLAN-B aggregate incoming messages within a configurable time frame before forwarding them as one larger packet. According to [9], this further increases the achievable throughput on the wireless medium.

In all of the aforementioned mechanisms, the spread of events across the MANET can be restricted by introducing hop counters and a Time to Live (TTL) for messages. However, this would limit the achievable completeness if events are to be disseminated to more remote areas of the network. Given our focus on location-based publish/subscribe, we instead apply the concept of geofencing to reduce load on the network as discussed in the following section.

4.3.1.3 Utilizing Geofencing for Location-Based Events

In our application scenario, location-based subscriptions always cover the physical proximity of a mobile client. Therefore, events published to a specific location do not necessarily need to be distributed across the whole MANET. Given that the size of the AoI is known (or at least has a well-known upper bound) the dissemination mechanism can utilize this knowledge to limit propagation of events to uninterested clients. This process is referred to as *geofencing*, meaning that clients check whether received messages are still relevant at their current locations before forwarding them.

In addition to basic geofencing, dedicated *geocasting* protocols for MANETs have been proposed [16]. They focus on forwarding messages to a geographic location or target area instead of a well-known IP address. We conducted a study on geocasting protocols for location-based services in an earlier work [7], showing the benefits of exploiting data and request locality within ad hoc dissemination mechanisms.

In BYPASS.KOM, we distinguish between local events and generic location-based events. This allows us to apply the concept of geofencing to our local dissemination mechanisms. A local event is only forwarded by a client if it matches one of the client's subscriptions. Assuming similar AoIs, events are only distributed locally within reasonably close proximity from their source.

However, this mechanism cannot be applied to generic location-based events. Here, more sophisticated geocasting protocols would need to be applied to reduce the spread of messages to areas where the information is not required. However, as we later discuss in the evaluation of BYPASS.KOM, distributing events over large distances in a pure ad hoc fashion is not feasible due to intermittent message loss and partitions of the network. Instead, we utilize transitions between local dissemination mechanisms as discussed in the following section. Executing transitions between dissemination mechanisms allows us to form islands of clients where events are distributed in an ad hoc fashion, as proposed in [19] and detailed in Sect. 4.4. With that, the concept of geofencing is also applied to generic location-based events.

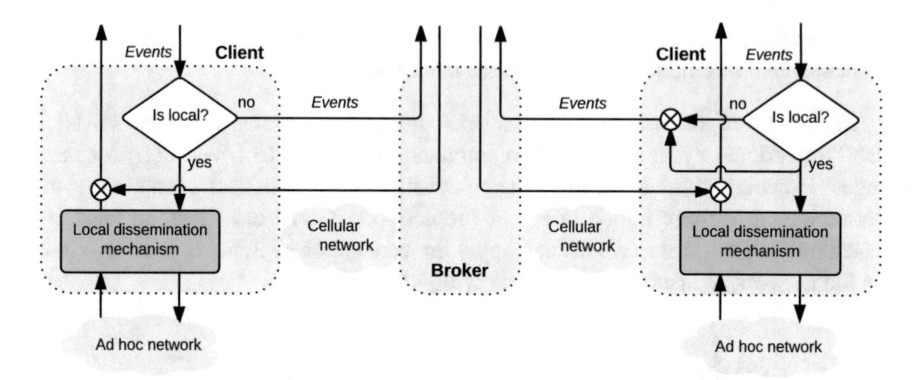

Fig. 4.8 Hybrid dissemination strategies: offloading (left) and augmentation (right)

4.3.2 Cloud-Based Hybrid Event Dissemination

As proposed in our previous works [19], we combine direct ad hoc dissemination of events with event dissemination via a cloud-based backend. The resulting hybrid system achieves low delivery delays for events that are distributed within vicinity of a client via ad hoc communication. At the same time, reliable event delivery can be realized via the cloud-based backend. We distinguish between two strategies, as illustrated in Fig. 4.8:

Cellular offloading. Clients only send events to the cloud which are not destined at a location within close proximity. From there on, the event is distributed by the cloud. All locally relevant events are only disseminated via an ad hoc dissemination mechanism. While this reduces the dissemination delay and lowers the utilization of the cellular connection, the successful delivery of local events cannot be guaranteed. Additionally, as the cloud does not receive all events at least once, a consistent global state cannot be guaranteed. This strategy is utilized by the client on the left side in Fig. 4.8.

Cloud augmentation. Clients send their events to the cloud where they are processed and then forwarded to all interested clients. Locally relevant events are additionally sent via an ad hoc dissemination mechanism with geofencing enabled, as discussed in the previous sections. Assuming a reliable cellular connection, as discussed beforehand, this strategy ensures event delivery. For local events, this substantially lowers the delay, as events are ideally received directly via the ad hoc dissemination mechanism. While this strategy does not lower the load on the cellular network, it guarantees that each event is delivered to interested clients at least once. The respective configuration is illustrated in Fig. 4.8 for the client on the right.

Combining the hybrid system design with transition-enabled dissemination mechanisms allows for central coordination of mechanism transitions, as presented in [19].

Furthermore, information available at the cloud-based broker helps in selecting suitable dissemination mechanisms. In the following section, we discuss how they are executed within BYPASS.KOM with and without central coordination.

4.3.3 Executing Transitions Between Local Dissemination Mechanisms

We consider two general models for the execution of transitions between different dissemination mechanisms: (i) central coordination and notification, and (ii) fully decentralized execution. The former requires a central entity, given by the cloud-based broker in hybrid dissemination models as originally proposed in [19]. The latter is required in fully decentralized settings, as discussed in [20] for the scenario of a large-scale music festival.

A centrally coordinated transition between two dissemination mechanisms behaves just like a total transition between different filter schemes, as discussed in Sect. 4.2.3. The coordinator (in our case: the broker) notifies each affected client of the transition, which is then executed locally. As for filter schemes, we refer to the respective transition between dissemination mechanism A and B as $T : A \rightarrow B$. In contrast to the transition between filter schemes, the broker itself does not execute a transition, as it does not utilize any ad hoc dissemination mechanism. As soon as a client receives the respective trigger, it executes the transition by interacting with the lifecycle of the respective mechanism proxy, as discussed in Sect. 5.2. We do not transfer any state information other than the set of currently registered listeners for incoming messages. This is due to the fact that the dissemination mechanisms utilized in our work do not construct complex topologies, as reasoned in Sect. 3.1.2.

In cases where we do not have a central entity, the trigger for a transition needs to be distributed to all clients via the ad hoc dissemination mechanism [20]. In that case, a decision might not reach all clients due to message loss or partitions within the network, leading to multiple dissemination mechanisms being used concurrently within the network. The same situation can arise for centrally coordinated transitions if the cellular connection is assumed to be unreliable or if the transmission delay varies substantially among clients.

To address this issue, we propose a self-healing mechanism for transition-enabled dissemination in BYPASS.KOM. This mechanism addresses two issues: (i) detecting a conflict and (ii) resolving the conflict. To be able to detect a conflict, all dissemination mechanisms within one transition-enabled component utilize the same port to listen for incoming messages. Additionally, each outgoing message contains a flag f_M identifying the mechanism M that created the message. Whenever a client receives a message that contains a flag which does not match the locally utilized dissemination mechanism, a conflict is detected. To provide this functionality in a mechanism-independent way, all outgoing and incoming messages are processed by a message proxy as already introduced for filter schemes in Sect. 4.2.2.

In order to resolve a detected conflict, additional information is required to decide which mechanism is to be used. Within our scenario, we assume that transitions are initiated by a single entity and not by individual clients. Therefore, we utilize a Lamport timestamp [15] by simply incrementing a counter whenever a transition is initiated. The value of this counter is contained within the respective transition execution plan distributed to clients. Clients simply piggyback their current value to outgoing messages. When receiving a message, clients simply compare their local value of the counter against the received value. In case the received value is greater than the local value, the transition $T : M_{\varnothing} \rightarrow M$ is being executed, with $M_{\varnothing}$ serving as generic placeholder matching any currently utilized mechanism, as further discussed in Sect. 4.4. The target mechanism M is identified by the flag f_M contained in the incoming message.

This self-healing mechanism maintains proper operation of the overall dissemination procedure by ensuring that clients eventually use the same mechanism. However, it cannot be applied to self-transitions (used to adapt parameters of a mechanism) or if transition decisions originate from multiple clients as result of a decentralized decision process. In such cases, more elaborate coordination strategies need to be deployed, as discussed in Sect. 5.2.3. We also consider using the self-healing mechanism as a lightweight alternative to global coordination to spread a transition decision within a group of clients, as later discussed in Sect. 4.4.

To further lower the load on the cellular infrastructure and, consequently, the associated cost for clients, we study the potential of gateways for locality-aware event dissemination in the following.

4.3.4 Locality-Aware Dissemination with Stateless Gateways

In each of the aforementioned hybrid dissemination strategies, the cloud-based broker still has to forward location-based events to a potentially large group of interested clients. Given the nature of location-based applications, these interested clients are usually within proximity of each other. Therefore, instead of delivering the event to each client, the broker can select a number of clients as gateways and instruct them to forward the event to their neighbors utilizing an ad hoc dissemination mechanism, as illustrated in Fig. 4.9. Therefore, the subscribers to an event are processed according to the currently utilized gateway selection algorithm as discussed in the following. The list of gateways and their assigned clients is then used to pack the outgoing events. If clients received a packed message with the respective instructions, it acts as gateway, as discussed in [23]. Besides notifying the local filter scheme, a transition-enabled local dissemination mechanism is used to distribute the event to assigned clients nearby.

This reduces the number of outgoing messages at the broker, lowering the load on the cellular network especially for densely populated regions. As discussed in [24], gateway selection algorithms can target different optimization goals, for example: (i) fair distribution of traffic and energy consumption of clients, (ii) maximum coverage

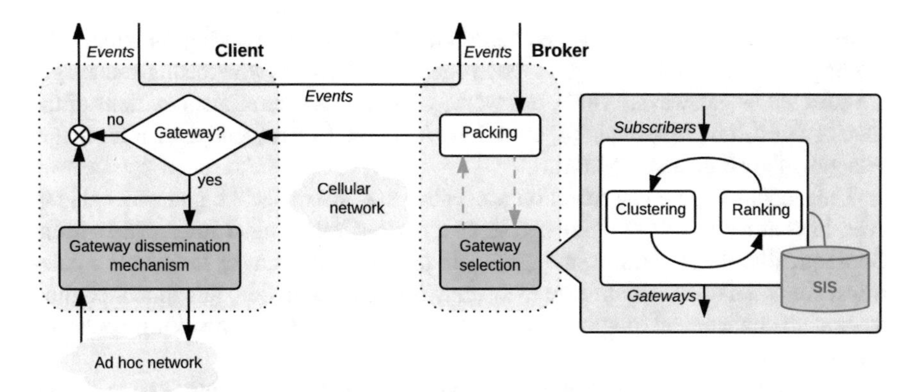

Fig. 4.9 Gateway-based locality-aware dissemination of events

with the minimal number of gateways, or (iii) longevity of selected gateways. We studied the impact of different gateway selection algorithms within the process of event dissemination in [23]. Given the dynamic nature of our scenario, we designed the role of a gateway in BYPASS.KOM to be stateless, with the broker entity managing state associated to the respective gateway selection algorithm. As a consequence, frequently switching gateways does not infer any additional cost in terms of messaging overhead. This is in stark contrast to existing approaches for gateway selection, as discussed in [23]. As gateways do not manage any state associated to the process of gateway selection, we can realize transitions between different gateway selection algorithms by exchanging the respective algorithm at the broker. Additionally, state information gathered and maintained by the broker can be transformed and transferred between different gateway selection algorithms.

More sophisticated gateway selection algorithms require additional information about mobile clients, such as the current battery state. Therefore, instead of integrating them into BYPASS.KOM, gateway selection algorithms are realized as standalone modules in a joint work with Nils Richerzhagen [24]. These modules retrieve all required information via the State Information System (SIS), which is later discussed in detail in Sect. 5.2.4. Part of this information can be fed into the SIS directly at the broker without additional communication cost; for example, the last known location of a mobile client. Being part of the SIMONSTRATOR.KOM platform, the SIS and its functionality is discussed in detail in Sect. 5.2.4.

We only consider gateway selection algorithms that operate solely on the last known location of a client, given that this information is available at the broker. Additional selection algorithms discussed in [24] could be utilized if the required state (e. g., battery status) is provided by an additional monitoring mechanism. Following the methodology proposed in [24], we distinguish between (i) a deterministic gateway selection followed by a clustering step to assign clients to gateways (DKC), (ii) a stochastic gateway selection followed by a clustering step (SKC), and (iii) an initial clustering step followed by a stochastic gateway selection within each cluster (CS).

In DKC, k gateways are selected based on a deterministic ranking function. The ranking function assigns a weight to each client. The k clients with the highest weight are selected as gateways. Within our work, we utilize a deterministic variant of the LEACH algorithm proposed in [8] as ranking function. Consequently, we refer to this selection algorithm as DKC-LEACH.

The stochastic gateway selection procedure SKC also selects k gateways. However, in contrast to DKC, k is treated as an expected value rather than a strict limit. Consequently, the stochastic strategy might assign more than (or less than) k gateways. We again utilize the LEACH algorithm as ranking function, this time including its random cluster-head rotation mechanism as defined in [8]. The resulting stochastic selection strategy is termed SKC-LEACH.

The aforementioned algorithms select gateways and subsequently assign clients to their nearest gateway according to a cluster function. The third strategy, CS, starts by clustering clients before selecting a single gateway within each cluster. We use the WCA algorithm [2] to determine a gateway within a cluster. WCA takes a client's location and its current velocity into account, as discussed in detail in [2]. The cluster-based gateway selection strategy relying on WCA is hereafter referred to as CS-WCA. Please refer to [24] for a detailed discussion of alternative selection algorithms and other ranking functions.

Regardless of the currently utilized algorithm, the gateway selection module is accessed via the API method `get_gateways(out_of, k)`, with k corresponding to the (expected) number of gateways as previously discussed. The variable `out_of` contains a list of client identifiers out of which gateways are to be selected. Based on the identifiers, additional state information (such as the location of the client) is retrieved by the respective selection algorithm using the State Information System. This is further discussed in Sect. 5.2.4. The method returns a mapping of selected gateways and their associated clients. This mapping is then utilized to bundle the corresponding events at the broker and forward them to the selected gateways, as discussed in [23]. Thereby, transition-enabled mechanisms for local event dissemination can be utilized in a hybrid dissemination mode.

Within BYPASS.KOM, we need to consider the dependencies between a local dissemination mechanism and the chosen gateway selection algorithm, as illustrated in Fig. 4.10. Considering one-hop dissemination, as studied in [23], gateways need to be selected such that all interested clients are within reach of direct communication.

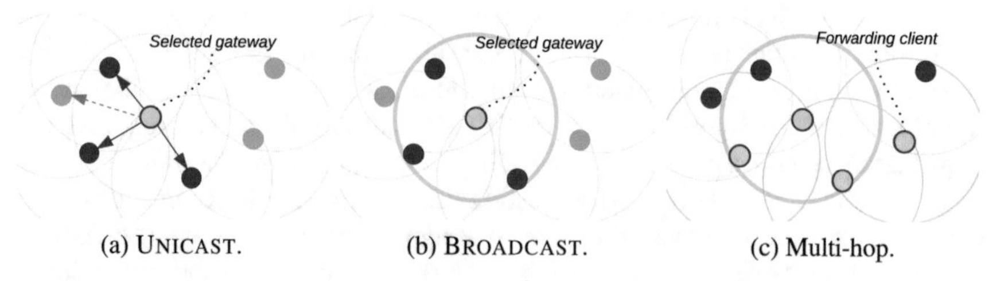

Fig. 4.10 Single-hop dissemination (UNICAST, BROADCAST) and multi-hop dissemination

Combined with multi-hop dissemination mechanisms as discussed previously, a single gateway might be sufficient to reach all interested clients in a target area. However, by selecting additional gateways that inject the event into the local ad hoc dissemination mechanism, the reliability of the overall dissemination process can be increased. We study the interdependencies between gateway selection and local dissemination within the scope of coexisting transition-enabled mechanisms, as discussed in the following section.

4.4 Coexistence of Transition-Enabled Mechanisms

To combine location-based filtering with locality-aware event dissemination, the respective mechanisms need to coexist in a publish/subscribe system. Additionally, transitions between these coexisting mechanisms need to be executed to adapt the system to application-specific client mobility and workload characteristics. In the following, we discuss the combined utilization of mechanisms for location-based filtering and locality-aware dissemination of events in BYPASS.KOM. We propose to adapt the choice of mechanisms to deal with application-specific client mobility by executing transitions near application-defined attraction points.

To this end, we assume that the locations and radii of attraction points are provided by the application, as previously discussed for the channel-based filter scheme ATTRACT in Sect. 4.2.4. As illustrated in Fig. 4.11, we consider two application-specific scenarios that require coexisting transition-enabled mechanisms. The scenarios resembles the hybrid dissemination strategies discussed in Sect. 4.3.2. However, the effects are limited to the area around attraction points and further combined with transitions between the utilized location-based filter schemes.

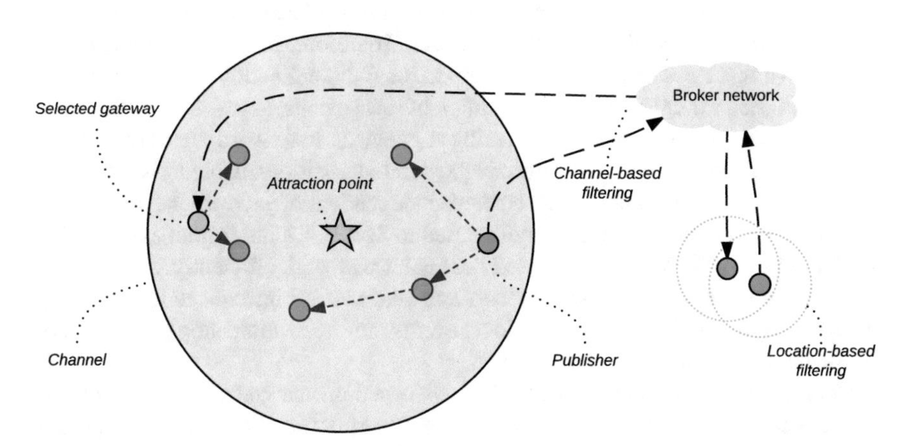

Fig. 4.11 Application-specific utilization of coexisting mechanisms

Cx-Adhoc. Ad hoc dissemination of local events is utilized in the area covered by an attraction point to reduce the delay of event delivery. In addition, a channel-based filter scheme is used for all clients in the area defined by an attraction point to avoid sending frequent context update messages. Clients not within proximity of an attraction point use a parametric filter scheme. Furthermore, they do not utilize an ad hoc dissemination mechanism.

Cx-Gateway. We utilize local dissemination mechanisms and filter schemes as discussed for CX-ADHOC. However, we additionally deploy stateless gateways to distribute events targeted at a location within a channel to all clients within vicinity of the attraction point. Instead of sending the event to each client individually, gateways combined with local dissemination are utilized to offload the cellular connection around attraction points.

In both cases, clients utilize a parametric filter scheme M_{pm} until they approach an attraction point. Within a circular area around attraction points, a channel-based filter scheme M_{ch} is used instead. The respective transition is initiated as soon as a client enters or leaves the area covered by an attraction point. As the transition only affects the client u entering (or leaving) the area, it is modeled as a partial transition $T^{\parallel}(u) : M_{\text{pm}} \rightarrow M_{\text{ch}}$ (or $T^{\parallel}(u) : M_{\text{ch}} \rightarrow M_{\text{pm}}$, respectively) as discussed in Sect. 4.2.3. Consequently, the client's state is migrated from one scheme's broker component to the other. In addition to switching the filter scheme, the client is instructed to activate a local ad hoc dissemination mechanism that is utilized whenever the client issues a local event. The dissemination mechanism is implicitly subject to geofencing, as ad hoc dissemination is deactivated once a client leaves the area of an attraction point.

For the case of CX-GATEWAY, we additionally utilize stateless gateways as previously discussed in Sect. 4.3.4. These gateways are utilized to disseminate events to all subscribers within a given channel, i. e., the clients in the area defined by an attraction point. The utilization of gateways requires a local dissemination scheme on both, gateways and the intended receivers. Therefore, in CX-GATEWAY, clients operate two instances of a transition-enabled local dissemination mechanisms. One instance is used for the ad hoc dissemination of local events, as previously discussed. The second instance is used by client acting as gateways to forward events received by the broker to nearby clients. By utilizing separate transition-enabled dissemination mechanisms, we can adapt the mechanisms independent of each other. This is motivated by the evaluation results later discussed in Sect. 6.4.2 and further supported by our findings in [23]. We show that stateless gateways work efficiently with one-hop dissemination mechanisms like UNICAST and BROADCAST introduced in Sect. 4.3.1, whereas the direct dissemination of local events requires a multi-hop dissemination mechanism to reach all interested clients.

We model the aforementioned transitions based on the generic concept of *execution plans* provided by the SIMONSTRATOR.KOM platform. Therefore, we present the resulting execution plans and concrete transitions when discussing our prototype of BYPASS.KOM in Sect. 5.3.2. We further discuss the adaptation of the respective execution plans to the currently observed client density around an attraction point.

With BYPASS.KOM, we introduce a framework that allows us to design and execute transitions between distinct mechanisms for location-based filtering and locality-aware event dissemination, respectively. We proposed distinct transition types and a common methodology for their execution. We focus on two perspectives: (i) individual mechanisms and their transitions and (ii) coexisting mechanisms with interdependent transitions. While BYPASS.KOM targets the domain of location-based publish/subscribe, we generalize core concepts and mechanisms in our second contribution, the SIMONSTRATOR.KOM platform. In the following chapter, we discuss the SIMONSTRATOR.KOM platform as an environment for the design and evaluation of transition-enabled communication systems.

References

1. Brimicombe A, Li Y (2006) Mobile space-time envelopes for location-based services. Trans GIS 10(1):5–23
2. Chatterjee M, Das SK, Turgut D (2002) WCA: a weighted clustering algorithm for mobile ad hoc networks. Clust Comput 5(2):193–204
3. Cugola G, Margara A, Migliavacca M (2009) Contextaware publish-subscribe: Model, implementation, and evaluation. In: Proceedings of IEEE symposium on computers and communications (ISCC). IEEE, pp 875–881
4. Eugster PTh, Garbinato B, Holzer A (2005) Location-based publish/subscribe. In: Proceedings of IEEE international symposium on network computing and applications. IEEE, pp 279–282
5. Friedman R, Shulman AK (2013) A density-driven publish subscribe service for mobile ad-hoc networks. Ad Hoc Netw 11(1):522–540
6. Fu Z, Meng X, Lu S (2002) How bad TCP can perform in mobile ad hoc networks. In: Proceedings of international symposium on computers and communications (ISCC). IEEE, pp 298–303
7. Groß C, Stingl D, Gottron C, Richerzhagen B, Münker C, Hausheer D (2012) Harnessing mobile ad hoc communication for decentralized location-based services. Technical report Peer-to- Peer Systems Engineering Lab, TU Darmstadt, Germany
8. Heinzelman WR, Chandrakasan A, Balakrishnan H (2000) Energy-efficient communication protocol for wireless microsensor networks. In: Proceedings of international conference on system sciences. IEEE, pp 10
9. Holzer A, Vessaz F, Pierre S, Garbinato B (2011) PLAN-B: proximity-based lightweight adaptive network broadcasting. In: Proceedings of international symposium on network computing and applications (NCA). IEEE, pp 265–270
10. Huang Y, Garcia-Molina H (2003) Publish/subscribe tree construction in wireless ad-hoc networks. In: Proceedings of international conference on mobile data management. Springer, pp 122–140
11. Ilarri S, Mena E, Illarramendi A (2010) Location-dependent query processing: Where we are and where we are heading. In: ACM Comput Surv (CSUR) 42(3):12
12. Jayaram KR, Eugster P, Jayalath C (2013) Parametric contentbased publish/subscribe. In: ACM Trans Comput Syst (TOCS) 31(2):4
13. Khelil A, Pedro José M, Christian B, Kurt R (2007) Hypergossiping: a generalized broadcast strategy for mobile ad hoc networks. Ad Hoc Netw 5(5):531–546
14. Knuth DE (1998) The art of computer programming: sorting and searching, vol 3. Pearson Education
15. Leslie L (1978) Time, clocks, and the ordering of events in a distributed system. Commun ACM 21(7):558–565

16. Maihofer C (2004) A survey of geocast routing protocols. In: IEEE Commun Surv Tutor 6(2)
17. Maling DH (2013) Coordinate systems and map projections. Elsevier
18. Pietzuch P, Eyers D, Kounev S, Shand B (2007) Towards a common API for publish/subscribe. In: Proceedings of ACM international conference on distributed event-based systems (DEBS). ACM, pp 152–157
19. Richerzhagen B, Stingl D, Hans R, Groß C, Steinmetz R (2014) Bypassing the cloud: peer-assisted event dissemination for augmented reality games. In: Proceedings of14th IEEE conference on peer-to- peer computing (P2P), pp 1–10 (Sept 2014)
20. Richerzhagen B, Wagener A, Richerzhagen N, Hark R, Steinmetz R (2016) A framework for publish/subscribe protocol transitions in mobile crowds. In: Proceedings of 10th international conference on autonomous infrastructure, management and security (AIMS). IFIP, pp 1–14 (June 2016)
21. Richerzhagen B, Wulfheide J, Koeppl H, Mauthe A, Nahrstedt K, Steinmetz R (2016) Enabling crowdsourced live event coverage with adaptive collaborative upload strategies. In: Proceedings of 17th international symposium on a world of wireless, mobile and multimedia networks (WoWMoM). IEEE, pp. 1–3
22. Richerzhagen B, Richerzhagen N, Zobel J, Schönherr S, Koldehofe B, Steinmetz R (2016) Seamless transitions between filter schemes for location-based mobile applications. In: Proceedings of 41st IEEE conference on local computer networks (LCN), pp 1–9 (Nov 2016)
23. Richerzhagen B, Richerzhagen N, Schönherr S, Hark R, Steinmetz R (2016) Stateless gateways-reducing cellular traffic for event distribution in mobile social applications. In: Proceedings of 25th international conference on computer communication and networks (ICCCN). IEEE, pp 1–9 (Aug 2016)
24. Richerzhagen N, Richerzhagen B, Walter M, Stingl D, Steinmetz R (2016) Buddies, not enemies: fairness and performance in cellular offloading. In: Proceedings of 17th international symposium on a world of wireless, mobile and multimedia networks (WoWMoM). IEEE, pp 1–9 (June 2016)
25. Van Brummelen G (2013) Heavenly mathematics: the forgotten art of spherical trigonometry. Princeton University Press
26. Werner M, Schwandke J, Hollick M, Hohlfeld O, Zimmermann T, Wehrle K (2016) STEAN: a storage and trans formation engine for advanced networking context. In: Proceedings of IFIP networking conference (IFIP Networking). IFIP, pp 91–99
27. Yoo S, Son JH, Kim MH (2009) A scalable publish/- subscribe system for large mobile ad hoc networks. J Syst Softw 82(7):1152–1162
28. Yu M, Guoliang L, Wang T, Feng J, Gong Z (2015) Efficient filtering algorithms for location-aware publish/subscribe. IEEE Trans Knowl Data Eng 27(4):950–963
29. Zhou Y, Xie X, Wang C, Gong Y, Ma W-Y (2005) Hybrid index structures for location-based web search. In: Proceedings of ACM international conference on information and knowledge management. ACM, pp 155–162

Chapter 5
SIMONSTRATOR.KOM: Platform for the Design and Evaluation of Mechanism Transitions

Insights obtained through domain-specific studies of mechanism transitions (as done with BYPASS.KOM) need to be consolidated and generalized towards a common transition methodology. This requires intense collaboration among researchers for the definition and later adoption of common design practices. With the SIMONSTRATOR.KOM platform [10, 21] as our second contribution, we support this process from the design of transition-enabled mechanisms to their evaluation in heterogeneous scenarios.[1] More specifically, the SIMONSTRATOR.KOM platform serves the following goals: (i) foster collaboration among researchers by supporting rapid prototyping and loose coupling among specific components, (ii) consolidate generalized design concepts for transition-enabled communication systems and their coordination, and (iii) support evaluations based on simulations and prototypical deployments.

We provide an overview of the platform in the following section, before discussing the transition-specific contributions in Sect. 5.2. These contributions deal with the design of transition-enabled mechanisms and their unified coordination through a *transition engine*. Finally, we discuss our prototype realization of BYPASS.KOM within the SIMONSTRATOR.KOM platform, serving as the foundation for the evaluation presented in Chap. 6.

5.1 Overview of the Platform

The Java-based SIMONSTRATOR.KOM platform consists of three main building blocks, as illustrated in Fig. 5.1: (i) a core *framework* for distributed communication systems, (ii) a number of *runtime environments* for execution and evaluation, and (iii) real-

[1]The SIMONSTRATOR.KOM platform, its runtime environments and the prototype of BYPASS.KOM described in this thesis are available online at www.simonstrator.com [Accessed March 8th, 2017].

B. Richerzhagen, *Mechanism Transitions in Publish/Subscribe Systems*,
Springer Theses, https://doi.org/10.1007/978-3-319-92570-7_5

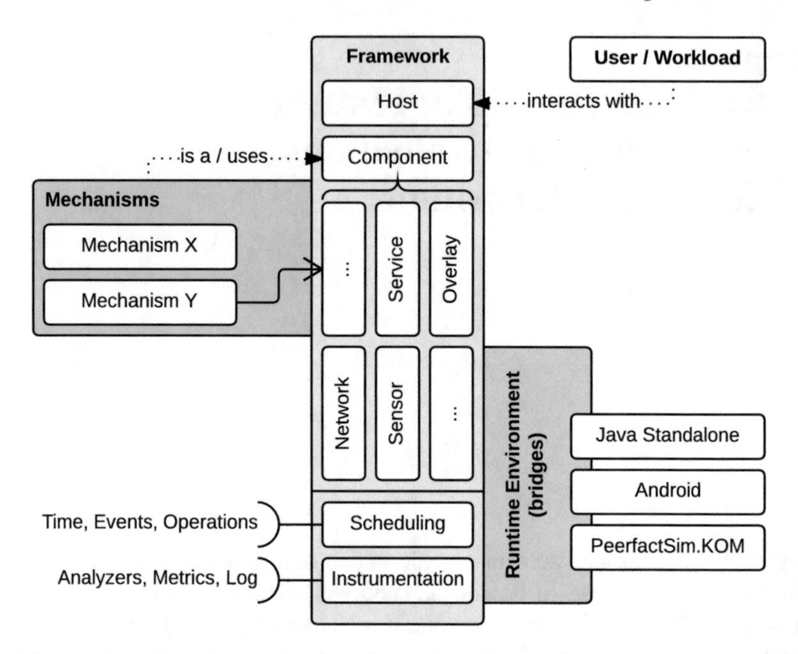

Fig. 5.1 Overview of the SIMONSTRATOR.KOM platform (adapted from [10])

izations of *specific mechanisms* within the framework. Realizations of mechanisms are usually separated into multiple projects, given that they represent individual researchers' current work. They are designed and implemented using the APIs provided within the core framework. This ensures interoperability between mechanisms and enables their execution in different runtime environments, as detailed in the following. An in-depth description of the platform and its components is provided in [10], albeit not focusing on transition-enabled mechanisms. In the following, we provide a brief summary of the platform as a foundation for the discussion of the transition-specific components and design concepts contributed in this thesis.

5.1.1 The Core Framework: Time, Events, and Instrumentation

The core SIMONSTRATOR.KOM framework consists of mechanism-specific APIs and a set of core components (or *global components*) providing abstractions for (i) scheduling of events, (ii) relative time calculations, and (iii) instrumentation. Abstractions for time and scheduling of events are required to support runtime environments that do not rely on continuous time, such as event-based simulators. Instrumentation is required to assess the performance of a mechanism, both in simulations and in real-world deployments. By using the provided abstractions, one can benefit from an extensive set of tools for collection and post-processing of data. Implementing a mechanism using the provided abstractions ensures that it can be executed on any of the runtime environments available within the SIMONSTRATOR.KOM platform, as later discussed in Sect. 5.1.3.

Scheduling a specific operation at a later point in time is a key functionality that is usually achieved in application threads through the use of sleep cycles. Within the SIMONSTRATOR.KOM, all scheduling is done through the `Event` wrapper provided by the framework:

```
void Event.schedule_with_delay(delay,
    event_handler, content, event_type).
```

The time until the event is executed, `delay`, is given in multiples of the time units provided by the `Time` wrapper, such as `Time.MINUTE` or `Time.SECOND`. Once the event is to be executed, the provided `event_handler` is notified by invoking

```
void event_occured(content, event_type).
```

The parameter `content` allows the user to pass an arbitrary object that is then passed to the respective event handler, as discussed hereafter. Similarly, the optional integer parameter `event_type` is passed to the event handler to enable easy distinction in case of multiple event types being processed by the same handler. For relative time calculations or absolute timestamps, rather than using the system clock, the current timestamp is obtained through the `Time.get_current_time()` method. Combined with the time units provided by the `Time` wrapper, this ensures identical behavior of all time-related calculations regardless of the utilized runtime environment.

Instrumentation aids in debugging during the design phase and in conducting measurements during subsequent evaluations. Within the SIMONSTRATOR.KOM platform, we offer pull- and push-based instrumentation in addition to simple logging functionality, as illustrated in Fig. 5.2. Pull-based instrumentation relies on a simple, predefined interface that enables access to specific internal states of a system via a *metric*. State can be accessed and observed through this interface in a simple and unified fashion, usually through periodic sampling. In contrast, push-based instrumentation allows the system to report specific actions or conditions on occurrence. The respective actions are usually mechanism-specific, requiring a custom instrumentation interface, referred to as *analyzer*.

Both methods of instrumentation are not expected to alter the internal state of the system that is being evaluated. Consequently, methods defined by a custom analyzer interface cannot return values. This is a fundamental requirement in the process of binding metrics and analyzers to a specific runtime environment. A runtime environment may include multiple analyzers implementing and binding to the same interface. This is transparently taken care of by the runtime environment, which multiplexes the respective method invocations to all analyzer instances. In case that there is no implementation of a specific analyzer available, the runtime environment automatically creates an empty stub.

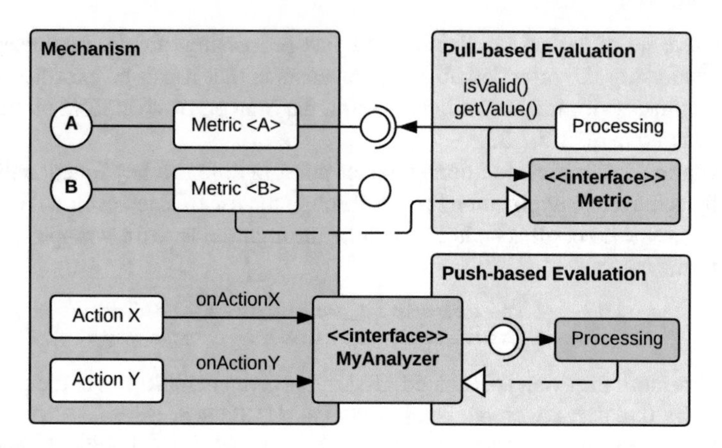

Fig. 5.2 Realization of push- and pull-based instrumentation (adapted from [10])

5.1.2 *Component-Based Composition of Hosts*

In addition to the aforementioned global components, mechanism-specific APIs for so-called *host components* are provided within the framework. A *host* describes an entity of the communication system, including mobile clients and intermediate network devices. The functionality of a host is defined by the set of available and configured host components. The SIMONSTRATOR.KOM framework offers APIs for a number of different protocols and services, including the API for location-based publish/subscribe proposed in Chap. 4. As indicated in Fig. 5.1, additional device-specific capabilities are offered as host components, including access to location sensors like GPS or available network interfaces for cellular and local ad hoc communication.

Host components can have multiple implementations, often depending on the runtime environment. Which implementation is to be used is configured within the respective runtime, with the individual components communicating with each other only through the APIs defined in the framework.[2] This enforces decoupling of functional dependencies and the actual realizations of the respective host components. Consequently, stub implementations of specific aspects of a communication system can easily be replaced with current research prototypes. Thereby, researchers can contribute to the big picture without creating too many dependencies on rapidly developing and possibly unstable research prototypes at the same time.

This is illustrated in Fig. 5.3 for the sensor used to determine a host's current location. The `LocationSensor` extends the generic `Sensor`, and applications interact with the respective interface. The actual realization, however, depends on the runtime environment and its configuration. On Android devices, for example, the device's actual GPS sensor can be utilized. In simulations, one might choose between a trace-based sensor implementation or a fully-fledged mobility model.

[2]For a detailed description and examples of configured hosts, the interested reader is referred to [10].

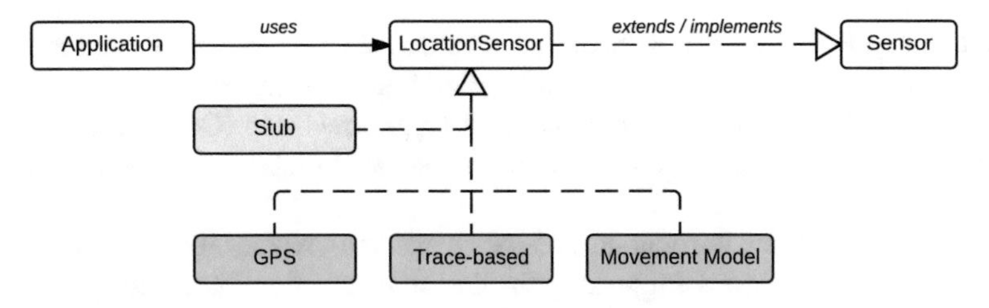

Fig. 5.3 Example of exchangeable component realization (adapted from [10])

The decoupling within the SIMONSTRATOR.KOM platform has proven to be vital for the Collaborative Research Centre "MAKI"during the development of demonstrators [9, 11, 13, 29]. Before going into detail on our transition-specific contributions to the SIMONSTRATOR.KOM platform, we briefly discuss the runtime environments used for the aforementioned demonstrators and the evaluation of BYPASS.KOM later presented in Chap. 6.

5.1.3 Runtime Environments

The SIMONSTRATOR.KOM platform includes a number of runtime environments for the execution and evaluation of systems developed within the framework. The current release 3.1 of the platform contains the following runtime environments:

PEERFACTSIM.KOM. An environment for event-based simulations relying on an extended version of the PEERFACTSIM.KOM overlay and network simulator [24]. This runtime environment was used for the evaluations presented in this thesis. The respective configuration and the utilized network and mobility models are described in more detail in Chap. 6.

Android. Given our focus on mobile applications, the Android runtime environment executes code on Android smartphones. It provides a basic stub for custom Android applications that can utilize systems and mechanisms implemented within our framework. The smartphone-based demonstrations published in [9, 11, 29] rely on this runtime environment.

Java Standalone. This environment enables the execution of systems on any computer with a Java runtime installed. This includes testbed environments [1] utilizing a headless mode and demonstration setups with custom user interfaces and visualizations.

It is important to note that the SIMONSTRATOR.KOM platform itself is *not* a simulator or emulator. Instead, it relies on a runtime environment to implement the respective core components for scheduling and time, as discussed previously and detailed in [10]. However, development of the SIMONSTRATOR.KOM platform is closely intertwined with the continued development of PEERFACTSIM.KOM. Consequently, the PEERFACTSIM.KOM runtime is the most feature-complete runtime.

Additionally, PEERFACTSIM.KOM has been extended with new models and tool integrations to specifically target the scenario of mobile social applications, as presented in [21, 25]. This includes the integration of OpenStreetMap (OSM) data for map-based navigation as foundation for more sophisticated mobility models as later discussed in Chap. 6.

Based on this brief introduction of the SIMONSTRATOR.KOM platform, we now discuss core contributions supporting the design of transition-enabled systems and the coordination of transitions.

5.2 Platform Support for Transitions

In our early contributions on transition-enabled mechanisms in the context of P2P video streaming [7, 28], the execution of transitions is realized as part of the streaming system. Consequently, encapsulation of the transition-enabled mechanisms relied on application-specific interfaces that cannot easily be applied to other application domains. Based on the design of BYPASS.KOM and our recent contributions on adaptive monitoring systems [17, 20], we extend and generalize the concept of transition-enabled mechanisms within the SIMONSTRATOR.KOM platform. The resulting abstractions for *transition-enabled mechanisms* and *mechanism proxies* are introduced in the following section. Utilizing these abstractions, we present an application and system agnostic *transition engine* that executes local transitions in Sect. 5.2.2. Subsequently, platform support for the coordination and execution of transitions that affect multiple devices based on *execution plans* is discussed in Sect. 5.2.3. Finally, we discuss the platform support for the monitoring, analysis, and planning steps of the MAPE-cycle in Sect. 5.2.4.

5.2.1 *Transition-Enabled Mechanisms and Mechanism Proxies*

The execution of transitions is a generic, mechanism-independent concept. Consequently, it has to be decoupled from the mechanism-specific APIs. In the SIMONSTRATOR.KOM platform, this is achieved by *mechanism proxies*, as illustrated in Fig. 5.4 and published in [5]. A mechanism proxy exposes the API of the original mechanism and transparently routes all method invocations to the currently active mechanism instance maintained within the proxy. When executing a transition, the transition engine interacts with the mechanism proxy to switch from one active mechanism to another, potentially executing a custom `AtomicTransition`, as illustrated. This is discussed in more detail in Sect. 5.2.2. For now, we focus on the mechanism proxies and their utilization.

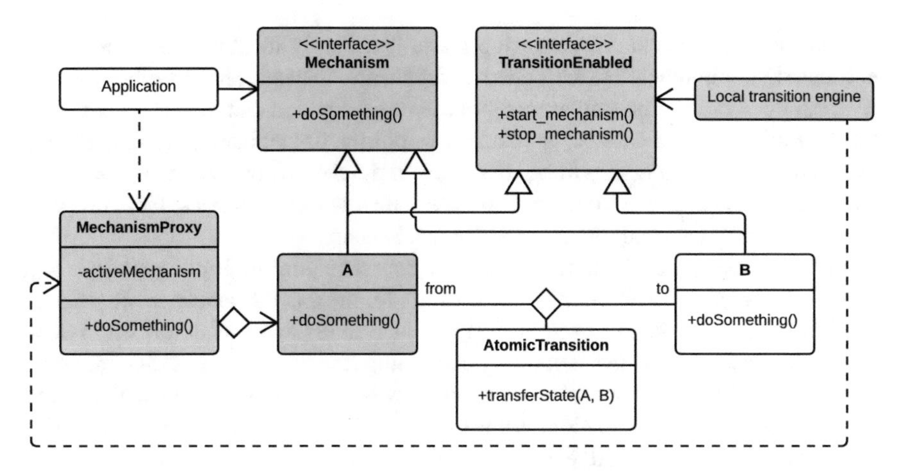

Fig. 5.4 Mechanism proxies within the SIMONSTRATOR.KOM platform

Proxies for transition-enabled mechanisms are created via the transition engine, a host component within the core framework of the SIMONSTRATOR.KOM. The transition engine offers the following method to create a new proxy for a given mechanism type, described by its application interface `C`:

```
Proxy create_mechanism_proxy(Class<C> api,
    default_instance, proxy_name);
```

The default mechanism instance provided when creating the proxy is used as active mechanism until any transitions are executed. The name of the proxy is required to address specific proxies when defining execution plans for transitions, as discussed later. Once the respective proxy is created, it can be utilized by the application just as any customary instance of the mechanism. Existing mechanism proxies can be accessed through the transition engine:

```
Proxy get_proxy(proxy_name, Class<C> api);
```

In case that a proxy is no longer required, the transition engine takes care of a proper shutdown of the currently active component.

If a transition is to be executed, the active mechanism within the respective mechanism proxy is altered or exchanged by the local transition engine. To this end, mechanisms have to implement two lifecycle methods defined by the `TransitionEnabled` interface: `start_mechanism` and `stop_mechanism`. In contrast to our earlier proposal in [5], we substantially simplified the lifecycle within the SIMONSTRATOR.KOM platform. In [5], the bootstrapping phase of a mechanism is part of the startup process. However, this definition makes it hard to define upper bounds for the time it takes before a mechanism has finished this process. Given that a transition can only be executed after the target mechanism has started, this leads to non-atomic behavior.

Within BYPASS.KOM and the lifecycle and transition model used by the SIMONSTRATOR.KOM platform, we follow a more radical approach to maintain atomicity of transitions. A mechanism's startup phase is a purely local operation and must not include communication with other components. Instead, only locally available state obtained during the state transfer phase of a transition can be used. Therefore, the respective method `start_mechanism` returns immediately, enabling atomic execution of transitions as will be discussed in the following section. If additional bootstrapping is required in cases where the state transfer does not suffice, this process is part of the normal operation of the mechanism.

In some cases, a communication system might require multiple instances of a specific mechanism proxy. This is, for example, the case if gateways are utilized in addition to direct local dissemination within BYPASS.KOM, as described in Sect. 4.4. In this case, one mechanism proxy containing transition-enabled dissemination mechanisms is used for direct dissemination of events, while a second instance is used for gateway-based dissemination. While the respective mechanism proxies can easily be distinguished by their name, the mechanisms themselves need to be designed (or at least configured) such that parallel operation is possible. In case of dissemination mechanisms, they need to use different transport layer ports, for example. This is further discussed in Sect. 5.3.1 for the prototype of BYPASS.KOM.

To support starting and stopping a transition-enabled mechanism, we introduce the *generic mechanism* $M_\varnothing$ as source or target mechanism of a transition. Executing a transition from $M_\varnothing$ to any other mechanism denotes a startup procedure, whereas a transition to $M_\varnothing$ shuts down a mechanism instance. Whenever a transition to $M_\varnothing$ is to be executed by a mechanism proxy, the respective proxy transparently provides a stub implementation of the proxy interface. This stub does not offer any functionality other than empty implementations of the methods defined in the interface. Thereby, it is ensured that invocations of the mechanism proxy still result in valid function calls. At the same time, the source mechanism of the transition is correctly terminated via its `stop_mechanism` lifecycle method just as during a normal transition.

If $M_\varnothing$ is specified as source mechanism of a transition (e. g., $T : M_\varnothing \rightarrow A$), it serves as a *wildcard* that matches any currently running mechanism. Thereby, transitions can be executed even if the currently running mechanism is not known to the coordinator, as discussed in the following section.

5.2.2 Transition Engine and Custom Transitions

As briefly mentioned in the previous section, the transition engine takes care of creating and maintaining proxies for transition-enabled mechanisms. Additionally, it is responsible for the local execution of transitions by interacting with the corresponding proxy. To execute a transition $T : A \rightarrow B$ from mechanism A to B, the following method needs to be triggered on the local transition engine:

```
void execute_atomic_transition(proxy_name,
   Class<C> target_class)
```

Note that the target mechanism's class is required, corresponding to the implementation of B. There is no need to specify A explicitly, as the transition to B is executed irregardless of the currently active mechanism of the proxy. Instead, the `proxy_name` is used to identify the correct mechanism proxy.

Next, the transition engine searches for an `AtomicTransition` object that matches the source and target mechanisms. An atomic transition contains code that is to be executed in order to transfer state between A and B. Custom realizations of an atomic transition can be registered at the transition engine:

```
void register_transition(proxy_name,
   AtomicTransition<A, B> transition)
```

Within each atomic transition, the method `transfer_state(`A, B`)` has to be implemented. While the application can only access the methods defined for the proxy upon creation, within `transfer_state(`A, B`)` one has full access to the individual functions defined by A and B. This, in turn, allows for arbitrary complex mechanism-specific state transformations. However, forcing developers to write explicit code for each potential transition would hinder the integration of new mechanisms and become tedious for larger choices of mechanisms. Therefore, the SIMONSTRATOR.KOM platform supports (i) fully automated state transfer using annotations and (ii) type inheritance for atomic transitions, as discussed in the following.

Fully automated state transfer relies on Java annotations. To enable automatic state transfer, fields within the respective mechanisms are annotated with `@TransferState`. This is, for example, used in BYPASS.KOM to transfer a reference to a location sensor between filter schemes:

```
@TransferState()
LocationSensor sensorReference;
```

During a transition, the local transition engine checks for matching annotations at the source and the target mechanism. If, in our example, both mechanisms contain an annotated field of the type `LocationSensor`, the value of the field is transferred to the new mechanism using the Java Reflection API. This corresponds to a custom `AtomicTransition` implementing the following `transfer_state` method:

```
void transfer_state(source, target) {
    target.sensorReference = source.
       sensorReference;
}
```

The annotation accepts an optional string value, e.g., `@TransferState ([name]")`. In this case, if the respective state is being transferred to the target mechanism of a transition, the local transition engine searches for a method `set[Name]` that accepts a variable of the respective type. This method is then invoked with the source mechanism's value. Thereby, a mechanism can actively react to transferred state by, for example, initializing an internal data structure. In case of automated state transfer, a *transformation* of the information is not possible. Consequently, source and target mechanism need to use the same state representation (data type). If this is not the case, a custom transition needs to be registered, as previously defined.

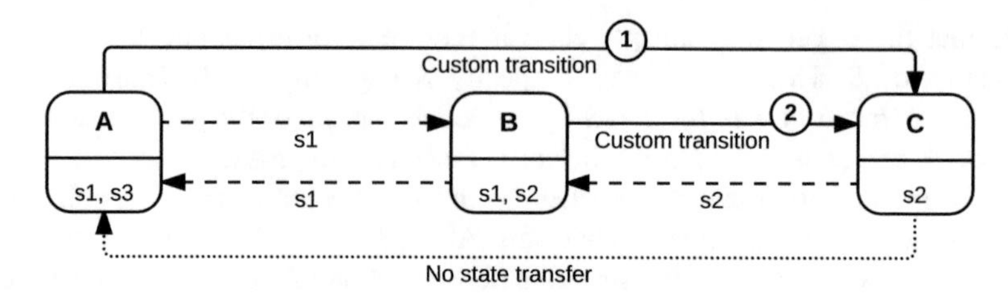

Fig. 5.5 Illustration of state transfer and type inheritance for (custom) transitions

When executing a transition, the local transition engine checks, whether a custom transition matches the specified source and target mechanisms. This process supports type inheritance, relying on the respective mechanisms provided by the Java programming language, as illustrated in Fig. 5.5. Here, three mechanisms (A, B, and C) are shown. Each mechanism contains one or more state variables s_i that are annotated with `@TransferState()`. Additionally, the custom transitions $T_1 : A \rightarrow C$ and $T_2 : M_{\varnothing} \rightarrow C$ have been registered at the local transition engine beforehand.

As shown in the figure, transition ① is executed using the custom implementation provided with T_1. The custom transition T_2 with $M_{\varnothing}$ as source mechanism also applies here, but the specification of T_2 is a closer match in terms of the type hierarchy of mechanisms ($M_{\varnothing}$ serves as parent for all mechanisms). The transition ②, however, is a match for the custom transition T_2. Consequently, instead of executing an automatically generated transition that would transfer state s_2, the custom transition is executed. In all other cases, matching state is automatically transferred as the respective state variables were properly annotated. Only for the transition from C to A, no matching state is found and, thus, no state information is transferred. The annotation-based automation of state transfer and the type inheritance for atomic transitions substantially reduce the effort required to integrate new mechanisms.

Irregardless of whether the transition from A to B uses a custom implementation of an `AtomicTransition` or whether it relies on the automated approach, it consists of the following steps, as illustrated in Fig. 5.6:

1. Create a new instance of mechanism B and transfer state from A to B.
2. Trigger `start_mechanism` at B (returns immediately in BYPASS.KOM).
3. Switch the active mechanism within the proxy to mechanism B.
4. Trigger `stop_mechanism` at A and discard the instance of mechanism A.

Steps 1 to 3 are considered a single atomic operation, if `start_mechanism` returns immediately, as discussed before.

Besides the execution of atomic transitions resulting from total or partial transitions within a system, we motivated the concept of self-transitions. Such transitions are used to alter the state of a mechanism without actually switching the mechanism instance. Self-transitions within the SIMONSTRATOR.KOM platform are also executed via the transition engine:

```
void execute_self_transition(proxy_name,
    SelfTransition<C> transition)
```

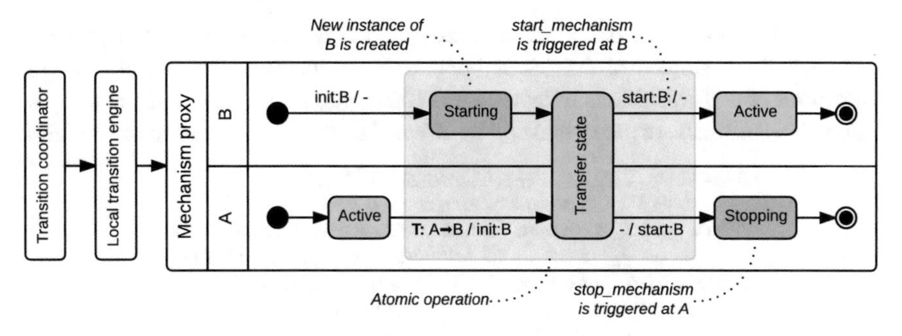

Fig. 5.6 Transition execution in the SIMONSTRATOR.KOM platform

Analogous to the functionality of an `AtomicTransition`, a `SelfTransition` realizing $T^{\bullet} : f(A)$ defines the method `alter_state(`A`)`, providing a reference to the currently active mechanism instance. Based on the methods available through the instance A, local state can be altered during the self-transition. As for atomic transitions, the SIMONSTRATOR.KOM platform includes annotations for mechanism state supporting the automated creation of matching self-transitions.

By annotating a field with `@MechanismState("[name]")`, the method `set[Name]` is invoked with the updated state variable whenever a self-transition is executed. This is especially useful for self-transitions that simply update a configuration parameter, as discussed in Sect. 4.2.5. If, for example, $T^{\bullet} : f(\text{GOSSIP}) = p \leftarrow 0.4$ is to be executed, and the state variable representing p in the GOSSIP dissemination mechanism is annotated with `@MechanismState("[Probability]")`, the following `SelfTransition` is automatically created and executed for the current mechanism instance of GOSSIP:

```
void alter_state(gossip) {
    gossip.setProbability(0.4);
}
```

The transition engine takes care of the local execution of transitions. This is only the last step in a chain of actions required to execute transitions across multiple hosts and multiple different mechanisms.

5.2.3 Coordination of Transitions on Multiple Hosts

Within the SIMONSTRATOR.KOM platform, a *transition coordinator* is responsible for the execution of transitions on multiple hosts. To this end, the transition coordinator supports the creation of an *execution plan*. The execution plan is an ordered sequence of transitions $EP = [T_1, \dots, T_n]$ that are to be executed, including any potentially required self-transitions. An execution plan is executed locally at the respective client; it does not involve any coordination. The execution plan serves as a blueprint that is distributed to all local transition engines that are involved in the global or partial transition. It can store an arbitrary number of *actions*, corresponding to either atomic transitions or self-transitions with associated state information. For each action, the name of the mechanism proxy that will be affected by the action needs to be provided:

```
plan = TransitionCoordinator.
    create_execution_plan();
plan.add_action(AtomicTranstionAction(
    proxy_name, Class<C> target_class));
plan.add_action(SelfTransitionAction(
    proxy_name, Class<C> target_class,
    state_identifier, state_value));
TransitionCoordinator.execute_plan(
    affected_hosts, plan);
```

With `execute_plan`, the plan is distributed to and executed on all hosts addressed by `affected_hosts`. Each host is equipped with an instance of a transition coordinator. In BYPASS.KOM, we utilize the cloud-based broker as a centralized entity that coordinates transitions. The transition coordinator component at mobile clients simply executes received plans by invoking the respective actions at the local transition engine. Fully distributed or hierarchically arranged coordinators can also be realized with the proposed structure, given that just the execution plan is predefined. The protocol used between coordinator instances is not defined by the platform and can be of arbitrary complexity, ranging from coordination via the global knowledge within a simulator to fully distributed consensus protocols for a decentralized realization of the MAPE-cycle [2]. In cases where the involved mechanisms cannot benefit from self-healing properties or a reliable network connection, more effort on the coordination side might be required.

5.2.4 Monitoring, Analysis, and Planning

So far, contributions focused on the execution of transitions as the last step of the MAPE-cycle introduced in Sect. 2.4. In the following, we briefly discuss contributions that support the monitoring, analysis, and planning phase of the cycle.

Providing data about the system's current state is an inevitable prerequisite for the analysis of the system's behavior as well as consequent planning and execution of transitions. Motivated by joint work on monitoring mechanisms [6, 16, 17, 20], we provide a local hub for data from multiple mechanisms within the SIMONSTRATOR.KOM platform. This State Information System, or SIS in short, is a local host component where mechanisms can register themselves as providers or consumers of specific data, based on the concept of monitoring access points proposed in [16]. The SIS interacts with the monitoring mechanism to distribute data across multiple hosts and resolve requests for data that is not locally available. Thereby, it provides data required by mechanisms such as the gateway selection algorithms presented in Sect. 4.3.4. Additionally, it provides a unified way for mechanisms in a single

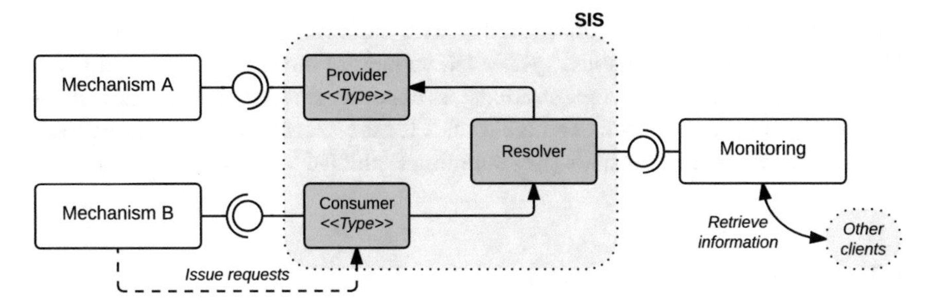

Fig. 5.7 Conceptual overview of the state information system

host to exchange data with each other. If, for example, a dissemination mechanism requires information about the local one-hop neighborhood of a mobile client, it can request that information from the SIS. The SIS tries to resolve the request locally by checking for providers (i. e., local mechanisms) that offer the respective data. When requesting information from the SIS, one can specify whether the SIS may utilize the monitoring mechanism to ask other clients.

Figure 5.7 provides a conceptual overview of the SIS and its interactions with information providers, consumers, and the monitoring mechanism. Besides supporting mechanisms during operation, the SIS can also be utilized to bootstrap mechanisms after a transition occurred. In addition to the direct transfer and transformation of state during a transition (as previously discussed), a mechanism can utilize the SIS to request state that is locally available at other mechanisms or resolved using the monitoring mechanism. The benefits resulting from this approach of bootstrapping mechanisms after a transition have been studied in the context of network routing protocols for MANETs in [27].

As is the case for all components within the SIMONSTRATOR.KOM platform, the actual realization of the SIS can differ depending on the runtime environment and its configuration. During early stages of mechanism design and the associated evaluations in a simulation environment, this allows us to rely on global knowledge provided by the simulator to single out other effects. A more realistic monitoring-based realization of the SIS can be used at any point in time without requiring modifications to the mechanisms utilizing it.

Information gathered by the monitoring mechanism and provided through the SIS is required during the analyzing and planning steps of the MAPE-cycle. We follow a pragmatic approach to analysis and planning, given that these phases of the MAPE-cycle are not in our core focus. In BYPASS.KOM, transitions are executed based on predefined rules, as presented in Sect. 4.4. All transitions are planned and coordinated by the cloud based broker entity based on locally available state information such as the current location or the number of subscribers.

Even though the analysis and planning steps are not within the focus of this thesis, other researches can benefit from concrete transition-enabled mechanisms available within the SIMONSTRATOR.KOM platform. More elaborated approaches to analysis and planning are explored in collaboration with other researches. This includes the description of transition-enabled systems in variability models to enable automatic derivation of valid system configurations, as done for BYPASS.KOM in [26]. Additionally, automated inference of suitable rules for the execution of transitions based on machine learning and genetic programming is studied in [3, 4].

5.3 Prototype of BYPASS.KOM

Within this section, we discuss the prototype of BYPASS.KOM realized based on the SIMONSTRATOR.KOM platform. We focus on the utilization of mechanism proxies, including parallel utilization of multiple instances and discuss how BYPASS.KOM is integrated into the MAPE-cycle for the coordination and execution of transitions.

5.3.1 Mechanism Proxies Utilized by Clients and Brokers

As discussed in Sect. 4, BYPASS.KOM is focused on two main functional components: filtering of events and local dissemination mechanisms including the utilization of gateways. To this end, each client utilizes five mechanism proxies, as described in the following and illustrated in Fig. 5.8.

Local Transport. This mechanism proxy provides transition-enabled access to a local transport interface and the underlying physical network technology. Thereby, dissemination mechanisms can operate regardless of the underlying network and transport technology. In BYPASS.KOM, the proxy provides mechanisms for UDP unicasts and broadcasts over Wi-Fi and Bluetooth. Within the evaluation, we rely on Wi-Fi functionality for its extended range. Bluetooth-based dissemination in BYPASS.KOM was studied prototypically in [11].

Direct Dissemination. Used for the direct ad hoc dissemination of local events, this mechanism proxy depends on the local transport proxy detailed before. It provides mechanisms for one hop and multi hop event dissemination and supports self healing as discussed in Sect. 4.3. During transitions from one dissemination mechanism to another, no state information needs to be transferred for the currently implemented protocols.

Client Gateway Logic. When acting as a gateway, a client needs to relay incoming events from the cloud to other nearby users. Relying on the stateless gateway approach published in [15] and previously discussed in Sect. 4.3.4, this proxy simply utilizes another instance of a proxy providing local dissemination mechanisms to forward messages to nearby clients.

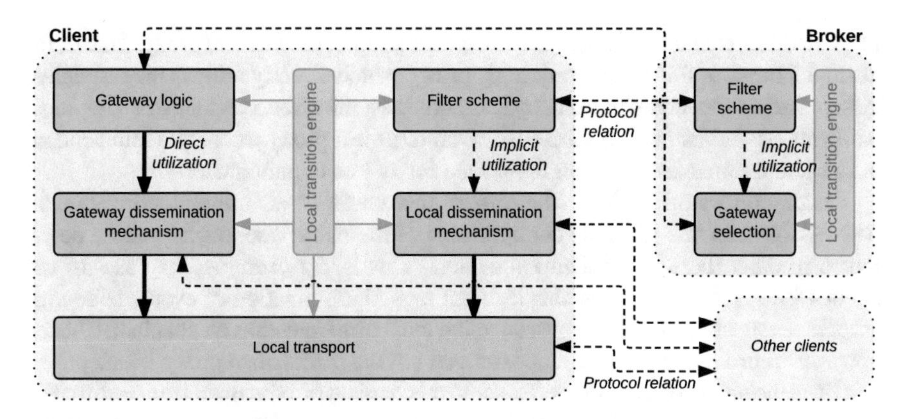

Fig. 5.8 Mechanism proxies in BYPASS.KOM and their dependencies

Gateway Dissemination. Functionally, this proxy is another instance of the direct dissemination proxy mentioned before. It operates on a separate port and is utilized solely by the client gateway logic to forward events to nearby clients. As such, it also relies on the local transport proxy to communicate via one of the available ad hoc network interfaces.

Filter Scheme. This proxy manages the client component implementation of the filter schemes introduced in Sect. 4.2.4. It provides the API for location-based publish/subscribe to the application. When executing a transition to another filter scheme, relevant state that needs to be transferred includes the current set of subscriptions and associated location requests. As filter schemes only communicate with their counterpart at the broker, they do not depend on any other mechanism proxy.

The broker component of BYPASS.KOM only utilizes two mechanism proxies: one for the utilized filter scheme and the other one for the utilized gateway selection algorithm.

Filter Scheme. This proxy manages the broker components of the location-based filter schemes discussed in Sect. 4.2.4. If total transitions are to be executed, it simply switches from one scheme to the other, taking care of state transfer and transformation. However, for partial transitions, multiple separate instances of proxies for filter scheme broker components are instantiated and managed.

Gateway Selection. This proxy acts as a wrapper for access to the generic framework of gateway selection algorithms provided in the SIMONSTRATOR.KOM based on the works in [15, 18]. It is used to execute transitions between different such algorithms during runtime. The interaction with the gateway selection framework is discussed in more detail in the following section.

The dependencies between mechanism proxies in BYPASS.KOM are illustrated in Fig. 5.8 and further modeled in [26]. Solid black arrows indicate a direct utilization of a transition-enabled mechanism from within another mechanism proxy. This is, for example, the case for the gateway logic proxy, which directly utilizes the respective dissemination mechanism proxy. Also, both dissemination mechanisms (for local publications and gateways) utilize the local transport proxy to operate independent of the actual transmission technology used for ad hoc communication.

Besides direct utilizations, there exist two cases of an implicit utilization in BYPASS.KOM. In this case, the configuration of the publish/subscribe system determines whether there is a utilization or not. This is, for example, the case for the client's filter scheme mechanism: if direct local dissemination of events is configured, the respective events are passed to the local dissemination mechanism. This is, however, coordinated by the framework and not the mechanism proxy itself.

Given the nature of a distributed system, mechanism proxies operating on different devices have a protocol relation with each other. This is, for example, the case for the client and broker component of a transition-enabled filter scheme, given that the components communicate via a custom context update protocol. Additionally, the gateway selection (or, more specifically, the packing component as introduced in Sect. 4.3.4) has a protocol relation to the client gateway logic, given that relevant information needs to be encoded in the respective messages. Similarly, the transport and dissemination mechanisms utilized by different mobile clients exhibit a protocol relation, as only matching mechanisms are able to successfully exchange messages. In the following, we discuss the coordination of the mechanism proxies introduced in this section in the context of coexisting mechanisms.

5.3.2 Coordination of Transitions for Coexisting Mechanisms

To combine location-based filtering and locality-aware dissemination of events we proposed the combined utilization of the respective transition-enabled mechanisms in Sect. 4.4. In the following, we discuss how the behavior defined in Sect. 4.4 is modeled and realized with *execution plans* and the *generic mechanism* $M_{\varnothing}$. Thereby, we show the applicability of the concepts and abstractions proposed in this chapter. Without loss of generality, we use concrete mechanisms for the discussion of the execution plans to ease understanding. The mechanisms and parameters introduced in this section are also used during the evaluation presented in Sect. 6.4.

First, we discuss the realization of the scenarios CX-ADHOCand CX-GATEWAY. We then propose and model the adaptation of execution plans depending on the observed client density around an attraction point, showing the flexibility of our approach to adaptive event brokering for mobile social applications.

5.3.2.1 Ad Hoc Event Dissemination Near Attraction Points

In addition to switching between filter schemes, we execute transitions to enable ad hoc dissemination of local events around attraction points. In the following, we present the resulting execution plans for the broker (EP_b) and the respective client

(EP_u). We use RADIAL as representative parametric filter scheme for clients outside the area covered by attraction points. We assign a channel to each attraction point by using ATTRACT as channel-based filter scheme for clients near attraction points.

When client u approaches an attraction point, a partial transition $T^{\parallel}(u)$: RADIAL → ATTRACT is executed by the broker. Thereby, the state associated to the client is transferred between the respective filter schemes' broker components. Subsequently, the total transition T : RADIAL → ATTRACT needs to be executed by the client. Additionally, a local dissemination scheme is activated. In this example, we use GOSSIP with a message forwarding probability of $p = 0.4$. Consequently, we execute the transition $T : M_{\varnothing} \rightarrow$ GOSSIP at the client, followed by a self-transition $T^{\bullet}$: f(GOSSIP), with f(GOSSIP) $= p \leftarrow 0.4$ setting the message forwarding probability. The impact of p on the performance of GOSSIP is further evaluated in A.3. The resulting execution plans for the broker and the client are as follows[3]:

$$EP_b = \left[T^{\parallel}(u) : \text{RADIAL} \rightarrow \text{ATTRACT}\right]$$

$$EP_u = \left[T : \text{RADIAL} \rightarrow \text{ATTRACT},\ T : M_{\varnothing} \rightarrow \text{GOSSIP},\ T^{\bullet} : f(\text{GOSSIP}) = p \leftarrow 0.4\right]$$

As soon as the client detects locally that it is about to leave the area covered by the assigned channel within ATTRACT, it sends a context update message to ATTRACT to request a new assignment. This is reported to the coordinator component, which initiates the respective execution plans to switch the client back to the RADIAL scheme. Additionally, the local dissemination mechanism is turned off:

$$EP_b = \left[T^{\parallel}(u) : \text{ATTRACT} \rightarrow \text{RADIAL}\right]$$

$$EP_u = \left[T : \text{ATTRACT} \rightarrow \text{RADIAL},\ T : \text{GOSSIP} \rightarrow M_{\varnothing}\right]$$

5.3.2.2 Gateway-Based Event Dissemination Near Attraction Points

Given that ATTRACT is a channel-based filter scheme, events targeted at a location within a channel in ATTRACT need to be distributed to all clients within vicinity of the attraction point. Instead of sending the event to each client individually, gateways combined with local dissemination can be utilized to offload the cellular connection, as proposed in the CX-GATEWAYscenario. However, the gateway selection algorithms discussed in Sect. 4.3.4 require knowledge about a client's current position at the broker. In ATTRACT, locations of clients within a channel are usually not updated, as the clients themselves detect whether they are about to leave the area covered by a channel. Therefore, we include an additional self-transition when switching the filter scheme: $T^{\bullet}$: f(ATTRACT) = enable_updates ← true. The

[3]In addition to activating an ad hoc dissemination protocol, we need to activate the respective communication interface at the mobile device, as also discussed for our Android prototype in [11]. We omit these transitions for brevity.

parameter `enable_updates` instructs the client component of ATTRACT to continue reporting location updates. We utilize the context update protocol of the filter scheme instead of a dedicated monitoring mechanism in our prototype and during evaluations, as the overhead in terms of message transmissions is negligible.

This configuration of BYPASS.KOM still disseminates local events via the GOSSIP mechanism as discussed in the previous section. To additionally enable dissemination of events from gateways to other clients nearby, we utilize a second instance of a transition-enabled dissemination mechanism. This is achieved by instantiating a second mechanism proxy that listens on a different port for incoming messages, as discussed in Sect. 5.3 for the prototype realization of BYPASS.KOM. As proposed in [15], gateways utilize the one hop dissemination mechanism BROADCAST. The dissemination mechanism is activated when a client enters a channel corresponding to an attraction point. This allows the client to listen to incoming messages sent by gateways nearby or to disseminate an event if selected as a gateway. Once the client leaves the area covered by the respective channel, it is instructed to deactivate the local dissemination mechanism as described earlier.

Given the fact that a client now uses two transition-enabled dissemination mechanisms in parallel, an execution plan needs to contain a reference as to which instance is affected by the transition. Therefore, we assign a name to each instance of a transition-enabled mechanism which is then used within the execution plan. This name corresponds to the name assigned to the mechanism proxy on creation, as discussed in Sect. 5.2.2. To apply a transition to a specific instance of a transition-enabled mechanism, we include the name of the instance. In the following, `gw` refers to the dissemination mechanism used by gateways and `lc` refers to the mechanism used for the direct dissemination of local events. Consequently, the following execution plan is sent to clients entering the channel of an attraction point:

$$EP_u = \Big[T : \text{RADIAL} \rightarrow \text{ATTRACT},\ T^{\bullet} : f(\text{ATTRACT}) = \text{enable\_updates} \leftarrow \text{true},$$
$$T_{\text{lc}} : M_{\varnothing} \rightarrow \text{GOSSIP},\ T^{\bullet}_{\text{lc}} : f(\text{GOSSIP}) = p \leftarrow 0.4,\ T_{\text{gw}} : M_{\varnothing} \rightarrow \text{BROADCAST}\Big]$$

When leaving the area covered by a channel, both local dissemination mechanisms are deactivated and the filter scheme is switched back to RADIAL:

$$EP_u = \Big[T : \text{ATTRACT} \rightarrow \text{RADIAL},\ T_{\text{lc}} : \text{GOSSIP} \rightarrow M_{\varnothing},\ T_{\text{gw}} : \text{BROADCAST} \rightarrow M_{\varnothing}\Big]$$

The execution plan at the broker does not differ from the previous case, as the broker only executes a partial transition of the filter scheme associated to the respective client. However, in contrast to the previous configuration, the broker needs to utilize the gateway selection algorithm when disseminating events to clients subscribed in the ATTRACT filter scheme. Therefore, the `get_gateways` method discussed in Sect. 4.3.4 is used to select gateways and associated clients out of the set of subscribers. The gateway selection requires a parameter k denoting the (expected) number of gateways, as previously discussed. We select k based on the number of clients subscribed to the respective channel within ATTRACT and depending on whether a one-hop or a multi-hop dissemination mechanism is used. This is discussed in detail during the evaluation presented in Sect. 6.4.2.

The number of clients and their density varies depending on the popularity and size of an attraction point and may change over time. In the following section, we therefore utilize transitions to continuously adapt the choice of mechanisms used nearby an attraction point based on the observed client density.

5.3.2.3 Rule-Based Adaptation of Utilized Mechanisms

In the previous configurations, transitions are only executed when clients enter or leave the area covered by a channel in ATTRACT. Once activated, the mechanisms utilized in the vicinity of an attraction point are not adapted to changing conditions during runtime. In this section, we discuss the execution of transitions to adapt the combination of mechanisms used near an attraction point based on the currently observed client density.

As previously discussed in Sect. 5.2.4, we focus on the execution of transitions rather than the reasoning on when to execute transitions. Therefore, we follow a simple rule-based approach for the execution of transitions, as discussed in the following. We select the set of mechanisms based on an assumed client density ρ. Whenever a client approaches an attraction point and subscribes to the respective channel in ATTRACT, the value of ρ is updated. The previous value of ρ is stored as ρ_{prev}.

Depending on ρ and ρ_{prev}, an execution plan for the adaptation of the current set of dissemination mechanisms is prepared by the broker. We refer to this execution plan as EP_a, with $EP_a = \left[T_{\text{lc}}\,,\ T_{\text{gw}}\right]$. In contrast to the previously discussed execution plans, the definitions of the transitions T_{lc} and T_{gw} now depend on ρ. We define T_{lc} such that PLAN- B is utilized for direct local dissemination if ρ falls below a threshold δ, and GOSSIP otherwise[4]:

$$T_{\text{lc}}: \begin{cases} M_{\varnothing} \rightarrow \text{PLAN- B} & \text{if } \rho \leq \delta \wedge \rho_{\text{prev}} > \delta \\ M_{\varnothing} \rightarrow \text{GOSSIP} & \text{if } \rho > \delta \wedge \rho_{\text{prev}} \leq \delta \\ \varnothing & \text{else.} \end{cases}$$

Note that instead of explicitly specifying the source mechanism, we use $M_{\varnothing}$ as generic source in both cases. Thereby, the transition can be executed regardless of the client's currently running mechanism. As discussed in Sect. 5.2.2, the transition engine automatically determines the most suitable transition realization depending on the transition specification and the locally utilized mechanism.

Analogous to the definition of T_{lc}, we adapt the dissemination mechanism used by gateways based on ρ:

[4]When switching to GOSSIP, the self-transition $T_{\text{lc}}^{\bullet} : f(\text{GOSSIP}) = p \leftarrow 0.4$ might need to be added to the execution plan. Otherwise, the new mechanism instance of GOSSIP will use its default probability.

$$T_{\text{gw}} : \begin{cases} M_{\varnothing} \to \text{UNICAST} & \text{if } \rho \leq \delta \wedge \rho_{\text{prev}} > \delta \\ M_{\varnothing} \to \text{BROADCAST} & \text{if } \rho > \delta \wedge \rho_{\text{prev}} \leq \delta \\ \varnothing & \text{else.} \end{cases}$$

We use UNICAST to disseminate events from gateways to clients for lower densities. As soon as the density exceeds ρ, BROADCAST is used instead. This is motivated by our evaluation results published in [15]. A unicast-based transmission benefits from higher bandwidth due to the modulation scheme chosen as a consequence of the Request to Send (RTS) and Clear to Send (CTS) "negotiation". Broadcasts rely on a more robust modulation scheme, thereby limiting the bandwidth of transmissions. However, as gateways need to send unicasts one after the other, the effective bandwidth decreases for increasing client density, as shown in [15].

The resulting execution plan EP_a is discarded, if it contains only empty transitions ($T_{\text{lc}} = \varnothing \wedge T_{\text{gw}} = \varnothing$). Otherwise, it needs to be executed by all subscribers U in the channel. As an alternative to notifying *all* subscribers using the transition coordinator, we also evaluate a strategy that relies on the self-healing capabilities of local dissemination mechanisms discussed in Sect. 4.3.3. Thereby, the execution plan only needs to be distributed to a single client, reducing the coordination overhead.

We evaluate the behavior resulting from coexisting transition-enabled mechanisms in Sect. 6.4. Our evaluation shows that the proposed execution plans enable the system to adapt to application-specific workload characteristics and client mobility.

The SIMONSTRATOR.KOM platform constitutes an important contribution towards generalizing the design of transition-enabled systems. Building on a rich foundation of domain-specific studies of transitions in video streaming [7, 22, 28], monitoring [17, 20], and publish/subscribe [8, 12, 14, 15], the abstractions provided by the platform support the design and evaluation of transition-enabled communication systems. Additionally, the clear separation of individual contributions into components fosters the re-use of existing mechanisms and benefits contributions in early stages of a project. The platform and its runtime environments form the basis for a growing number of demonstrators [9, 11, 13, 19, 23, 29], including demonstrations of dissemination mechanism transitions [11] and filter scheme transitions [13] in BYPASS.KOM.

References

1. Berman M, Demeester P, Lee JW, Nagaraja K, Zink M, Colle D, Krishnappa DK, Raychaudhuri D, Schulzrinne H, Seskar I et al (2015) Future internets escape the simulator. Commun ACM 58(6):78–89
2. Dolev D, Dwork C, Stockmeyer L (1987) On the minimal synchronism needed for distributed consensus. J ACM (JACM) 34(1):77–97
3. Frömmgen A, Rehner R, Lehn M, Buchmann A (2015) Fossa: learning ECA rules for adaptive distributed systems. In: Proceedings IEEE international conference on autonomic computing (ICAC). IEEE, pp 207–210

4. Frömmgen A, Rehner R, Lehn M, Buchmann A (2015) Fossa: using genetic programming to learn ECA rules for adaptive networking applications. In: Proceedings IEEE conference on local computer networks (LCN). IEEE, pp 197–200
5. Frömmgen A, Richerzhagen B, Röckert J, Hausheer D, Steinmetz R, Buchmann A (2015) Towards the description and execution of transitions in networked systems. In: Proceedings 9th international conference on autonomous infrastructure, management and security (AIMS). IFIP, pp 17–29
6. Hark R, Richerzhagen N, Richerzhagen B, Rizk A, Steinmetz R (2017) Towards an adaptive selection of loss estimation techniques in software-defined networks. In: Proceedings IFIP networking conference (IFIP Networking). IEEE, pp 1–9
7. Richerzhagen B (2012) Supporting transitions in peer-to-peer video streaming. In: Master's thesis, Technische Universität Darmstadt
8. Richerzhagen B, Stingl D, Hans R, Groß C, Steinmetz R (2014) Bypassing the cloud: peer-assisted event dissemination for augmented reality games. In: Proceedings 14th IEEE conference on peer-to- peer computing (P2P), pp 1–10
9. Richerzhagen B, Wilk S, Rückert J, Stohr D, Effelsberg W (2014) Transitions in live video streaming services. In: Proceedings workshop on design, quality and deployment of adaptive video streaming (VideoNEXT). ACM, pp 37–38
10. Richerzhagen B, Stingl D, Rückert J, Steinmetz R (2015) Simonstrator: simulation and prototyping platform for distributed mobile applications. In: Proceedings 8th international conference on simulation tools and techniques (SIMUTOOLS). ACM, pp 99–108
11. Richerzhagen B, Schiller M, Lehn M, Lapiner D, Steinmetz R (2015) Transition-enabled event dissemination for pervasive mobile multiplayer games. In: Proceedings 16th international symposium on a world of wireless, mobile and multimedia networks (WoWMoM). IEEE
12. Richerzhagen B, Wagener A, Richerzhagen N, Hark R, Steinmetz R (2016) A framework for publish/subscribe protocol transitions in mobile crowds. In: Proceedings 10th international conference on autonomous infrastructure, management and security (AIMS). IFIP, pp 1–14
13. Richerzhagen B, Richerzhagen N, Zobel J, Schönherr S, Koldehofe B, Steinmetz R (2016) Demo: seamless transitions between filter schemes for location-based mobile applications. In: Demonstrations of the 41st IEEE conference on local computer networks (LCNDemos), pp 1–3
14. Richerzhagen B, Richerzhagen N, Zobel J, Schönherr S, Koldehofe B, Steinmetz R (2016) Seamless transitions between filter schemes for location-based mobile applications. In: Proceedings 41st IEEE conference on local computer networks (LCN), pp 1–9
15. Richerzhagen B, Richerzhagen N, Schönherr S, Hark R, Steinmetz R (2016) Stateless gateways—reducing cellular traffic for event distribution in mobile social applications. In: Proceedings 25th international conference on computer communication and networks (ICCCN). IEEE, pp 1–9
16. Richerzhagen N, Li T, Stingl D, Richerzhagen B, Steinmetz R, Santini S (2015) A step towards a protocol-independent measurement framework for dynamic networks. In: Proceedings 40th IEEE conference on local computer networks (LCN). IEEE, pp 462–465
17. Richerzhagen N, Stingl D, Richerzhagen B, Mauthe A, Steinmetz R (2015) Adaptive monitoring for mobile networks in challenging environments. In: Proceedings 24th international conference on computer communication and networks (ICCCN). IEEE, pp 1–8
18. Richerzhagen N, Richerzhagen B, Walter M, Stingl D, Steinmetz R (2016) Buddies, not enemies: fairness and performance in cellular offloading. In: Proceedings 17th international symposium on a world of wireless, mobile and multimedia networks (WoWMoM). IEEE, pp 1–9
19. Richerzhagen N, Richerzhagen B, Lipinski M, Weckesser M, Kluge R, Hark R, Steinmetz R (2016) Exploring transitions in mobile network monitoring in highly dynamic environments. In: Demonstrations of the 41st IEEE conference on local computer networks (LCN-Demos), pp 1–3

20. Richerzhagen N, Richerzhagen B, Hark R, Stingl D, Steinmetz R (2016) Limiting the footprint of monitoring in dynamic scenarios through multi-dimensional offloading. In: Proceedings 25th international conference on computer communication and networks (ICCCN). IEEE, pp 1–9
21. Richerzhagen N, Richerzhagen B, Stingl D, Steinmetz R (2017) The human factor: a simulation environment for networked mo bile social applications. In: Proceedings international conference on networked systems (NetSys). IEEE, pp 1–8
22. Rückert J, Richerzhagen B, Lidanski E, Steinmetz R, Hausheer D (2015) TopT: supporting flash crowd events in hybrid overlay-based live streaming. In: Proceedings 14th IFIP networking conference (Networking). IEEE, pp 1–9
23. Stein M, Kluge R, Mirizzi D, Wilk S, Schürr A, Mühlhäuser M (2016) Transitions on multiple layers for scalable, energy efficient and robust wireless video streaming. In: Proceedings international conference on pervasive computing and communication workshops (PerCom Workshops). IEEE, pp 1–3
24. Stingl D, Gross C, Rückert J, Nobach L, Kovacevic A, Steinmetz R (2011) Peerfactsim. kom: a simulation framework for peer-to-peer systems. In: Proceedings international conference on high performance computing and simulation (HPCS). IEEE, pp 577–584
25. Stingl D, Richerzhagen B, Zöllner F, Gross C, Steinmetz R (2013) PeerfactSim.KOM: take it back to the streets. In: Proceedings international conference on high performance computing and simulation (HPCS). IEEE, pp 80–86
26. Weckesser M, Lochau M, Schnabel T, Richerzhagen B, Schürr A (2016) Mind the gap! automated anomaly detection for potentially unbounded cardinality-based feature models. In: Fundamental approaches to software engineering. Springer, Heidelberg, pp 158–175
27. Werner M, Schwandke J, Hollick M, Hohlfeld O, Zimmermann T, Wehrle K (2016) STEAN: a storage and transformation engine for advanced networking context. In: Proceedings IFIP networking conference (IFIP Networking). IFIP, pp 91–99
28. Wichtlhuber M, Richerzhagen B, Rückert J, Hausheer D (2014) TRANSIT: supporting transitions in peer-to-peer live video streaming. In: Proceedings IFIP networking conference (IFIP Networking). IEEE, pp 1–9
29. Wilk S, Rückert J, Stohr D, Richerzhagen B, Effelsberg W (2015) Efficient video streaming through seamless transitions between Unicast and broadcast. In: Proceedings 2nd conference on networked systems (NetSys), pp 1–2

Chapter 6
Evaluation of Mechanism Transitions

Based on our prototype of BYPASS.KOM within the SIMONSTRATOR.KOM platform, we conduct an extensive evaluation of mechanism transitions in different aspects of publish/subscribe systems. Furthermore, we highlight the impact of key design decisions such as state transfer mechanisms and self-healing properties on the performance of the system during transitions. We detail our evaluation goals and methodology in the following, based on which we discuss our evaluation setup in Sect. 6.1. The setup includes the map-based mobility models utilized to capture real-world behavior of mobile clients as discussed in Sect. 6.1.1 and the corresponding application workload models presented in Sect. 6.1.2.

The evaluation addresses the mechanisms integrated into BYPASS.KOM and their transitions as presented in Sect. 6.2. Consequently, we evaluate the impact of transitions for location-based filtering in Sect. 6.2, followed by the evaluation of transition-enabled mechanisms for locality-aware event dissemination in Sect. 6.3. The combination of location-based filtering and locality-aware dissemination and the resulting coexistence of transition-enabled mechanisms is evaluated in Sect. 6.4.

The goal of this evaluation is to characterize mechanism transitions as a means to adapt a communication system within the application domain of location-based publish/subscribe. This includes understanding the impact of a transition execution on the system's performance during runtime. Given our research goal, we evaluate how transitions contribute to combining and composing mechanisms for location-based filtering and locality-aware dissemination in a highly adaptive fashion. In addition to the evaluation of concrete transitions between filter schemes and dissemination mechanisms as proposed in Chap. 4, we also assess the applicability of our generic design concepts for transition-enabled systems proposed in Chap. 5.

To characterize and isolate the impact of a transition execution, we need to compare the transition-enabled system against static mechanisms in a reproducible setup.

B. Richerzhagen, *Mechanism Transitions in Publish/Subscribe Systems*,
Springer Theses, https://doi.org/10.1007/978-3-319-92570-7_6

At the same time, the setup needs to reflect the dynamics and heterogeneity of mobile social applications, as previously discussed. Achieving a reproducible setup in real-world measurements is hard to achieve, especially at the scale required to assess the dynamics of mobile social applications. Therefore, we rely on event-based simulations of our proposed system in the SIMONSTRATOR.KOM platform, utilizing models for client mobility derived from earlier studies [3, 22, 31]. The evaluation setup and the utilized models are further discussed in the following section.

6.1 Evaluation Setup

We evaluate our proposed mechanisms and their transitions using the prototype of BYPASS.KOM implemented in the SIMONSTRATOR.KOM platform as described in Sect. 5.3. We use an updated version [26] of the event-based simulator PEERFACTSIM.KOM [29] as underlying simulation engine. The updated version contains additional mobility models specifically targeted at our application scenario, as detailed in the following section. To evaluate the publish/subscribe system under realistic workloads, we propose workload models that capture key communication characteristics of our scenario. The workload models are described in Sect. 6.1.2. The network model utilized for our evaluation is detailed in Sect. 6.1.3. Finally, the evaluation metrics used to capture the behavior of transition-enabled mechanisms are introduced in Sect. 6.1.4, including a brief introduction of the measurement methodology and the plot types used in this chapter.

6.1.1 Mobility Models

Given the scenario of mobile social applications, human mobility needs to be accurately modeled in our evaluation [5, 6]. To this end, we extended our earlier work on modeling human mobility [26, 29, 31]. By integrating the corresponding mobility models into the SIMONSTRATOR.KOM platform, we evaluate our proposed transition-enabled publish/subscribe system under realistic client mobility, employing real-world map data. For comparison, we also conduct evaluations with the generic Random Waypoint Model [10]. This supports comparing our results to other state of the art approaches that still rely on simple mobility models [15]. Additionally, it provides insights on how an accurate mobility model exposes important characteristics of a communication system that would be left unnoticed in a generic scenario. The mobility models discussed in this section are summarized in Table 6.1. For each mobility model, we describe the *local movement* strategy (i. e., the model used to approach a target location) and the strategy used to select a *target location* in the following.

Table 6.1 Mobility models and their characteristics

Mobility model	Local movement	Target location
DA	Map-based, Darmstadt	Attraction points Darmstadt
MA	Map-based, Mannheim	Attraction points Mannheim
DA-RND	Map-based, Darmstadt	Random location
MA-RND	Map-based, Mannheim	Random location
RWP	Linear	Random location

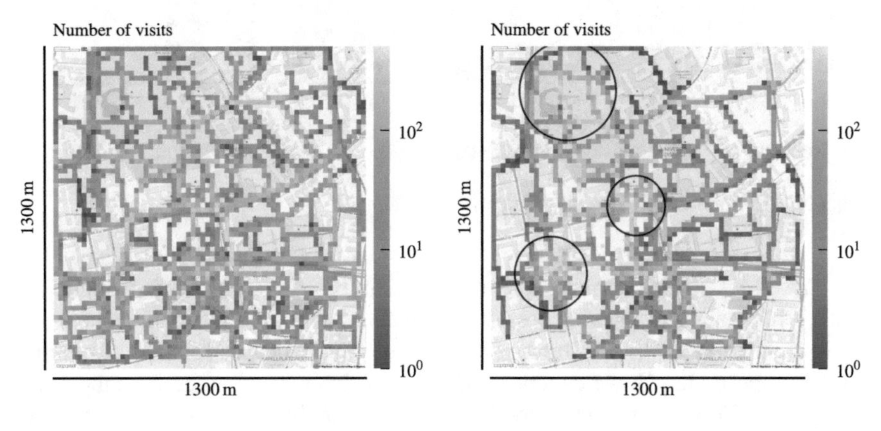

(a) Random target destinations (DA-RND). (b) Attraction points (DA).

Fig. 6.1 Client mobility for the city center of Darmstadt

6.1.1.1 Map-Based Mobility

Based on our earlier work [26, 29], we rely on real-world map data to model client mobility. We utilize OpenStreetMap (OSM) as map data provider, and the GraphHopper open-source library[1] for navigation. Relying on the routing capabilities of GraphHopper, clients follow pedestrian walkways to reach their target destination. The target destination is either selected randomly or among the set of attraction points defined by the application. As soon as a client reaches its target destination, the client pauses for a random time t_{p}, randomly chosen from the uniformly distributed interval between $t_{\mathrm{p,min}}$ and $t_{\mathrm{p,max}}$, before determining the next target location.

The resulting mobility characteristics are shown in Fig. 6.1a. The plots show the accumulated visits of clients over a time period of 30 min in the simulated area of $1,300 \times 1,300$ m in the city center of Darmstadt, Germany. Visits are aggregated cells of 20×20 m. Client mobility is limited to pedestrian walkways, as shown in Fig. 6.1a. For random target destinations, the number of visits is spread more evenly across the simulated area, including visits to more remote areas. If attraction points

[1] www.graphhopper.com/open-source/ [Accessed March 8th, 2017].

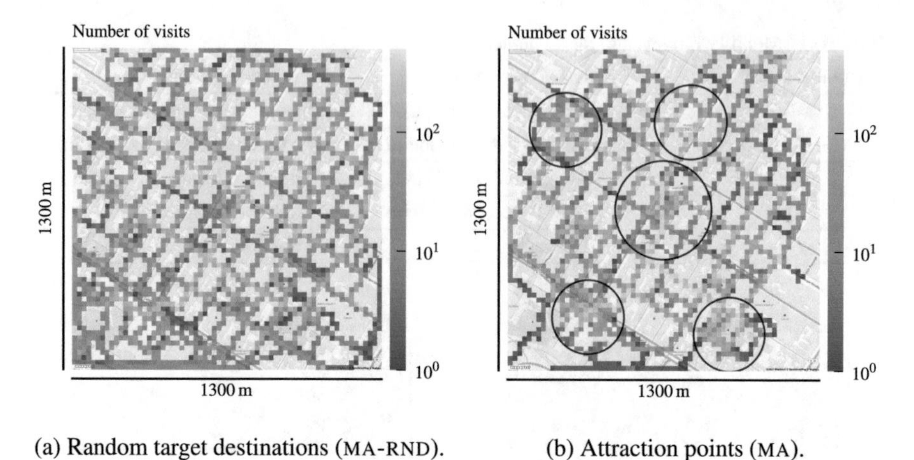

(a) Random target destinations (MA-RND). (b) Attraction points (MA).

Fig. 6.2 Client mobility for the city center of Mannheim

are chosen instead, the number of visits in vicinity of attraction points increases, while remote areas of the map are not visited at all, as shown in Fig. 6.1b.

To capture the essence of more grid-like mobility models such as the Manhattan model proposed in [2], we conduct additional evaluations based on the map of the city center of Mannheim, Germany. The resulting mobility characteristics are shown in Fig. 6.2a, b clearly highlighting the grid-like nature of Mannheim's city center. In comparison to Darmstadt, visits are more equally distributed across the simulated area. This is due to the fact that within the grid-like structure of Mannheim, the GraphHopper navigation library tends to find multiple different routes with similar length for a target location.

For both cities, attraction points are placed at public parks according to OSM data. All attraction points have an equal probability of being selected as a client's target location, referred to as the attraction point's *weight*. To prevent clients from targeting the exact same locations, we add a random offset from the center location of the attraction point. This offset is modeled via two Gaussian distributions with $\mathcal{N}_x\left(x_0, \sigma^2\right)$ and $\mathcal{N}_y\left(y_0, \sigma^2\right)$ that are used to draw the x and y coordinate around the attraction point (x_0, y_0). To ensure that more than 99% of the resulting target locations are within a radius r around the current attraction point, we set $\sigma = r/3$. For the attraction points generated from OSM data, we assume a default radius of $r = 25\,\text{m}$ to ensure that clients approach slightly different target coordinates.

To capture the real-world behavior of clients in augmented reality applications accurately, we manually assign higher probabilities and larger radii to a subset of the attraction points. Consequently, a greater number of clients visit the vicinity of these attraction points. To this end, we utilize data on in-game attraction points that is openly available for the augmented reality games Ingress and Pokémon Go. For Ingress, the location of *Portals* and their significance is available on Ingress Intel.[2]

[2]www.ingress.com/intel, Google account required [Accessed March 8th, 2017].

The location of in-game *Gyms* in Pokémon Go is gathered and provided by PokemonGoMap.[3] Both data sources motivate our choice of manually weighted attraction points for Darmstadt and Mannheim. Given the characteristics of the respective applications, these attraction points coincide with pedestrian zones and public parks. These manually altered attraction points are shown in Figs. 6.1b, 6.2b as black circles indicating their radius.

6.1.1.2 Random Waypoint Mobility as Baseline

To assess the impact of the mobility models in comparison to a baseline frequently used in related works, we include the Random Waypoint Model [10] in our evaluation. Here, clients select a random target location and approach that location following a straight line. After reaching their target location, they pause for a time t_p before continuing to a new randomly chosen target location. As for the map-based models, t_p is uniformly distributed between $t_{p,min}$ and $t_{p,max}$.

The resulting mobility characteristics are shown in Fig. 6.3a. The visits are distributed more evenly across the simulated area, with areas in the center being visited more often than the outskirts. Target locations with longer pause times are also clearly visible. Figure 6.3b shows a histogram of the number of times a cell is visited for the map-based models of Darmstadt (DA) and Mannheim (MA) and the Random Waypoint Model (RWP). In comparison to the map-based models, the visits are more evenly distributed and there are no locations with more than 200 visits. However, the nature of human mobility (especially in an urban scenario) is not accurately captured

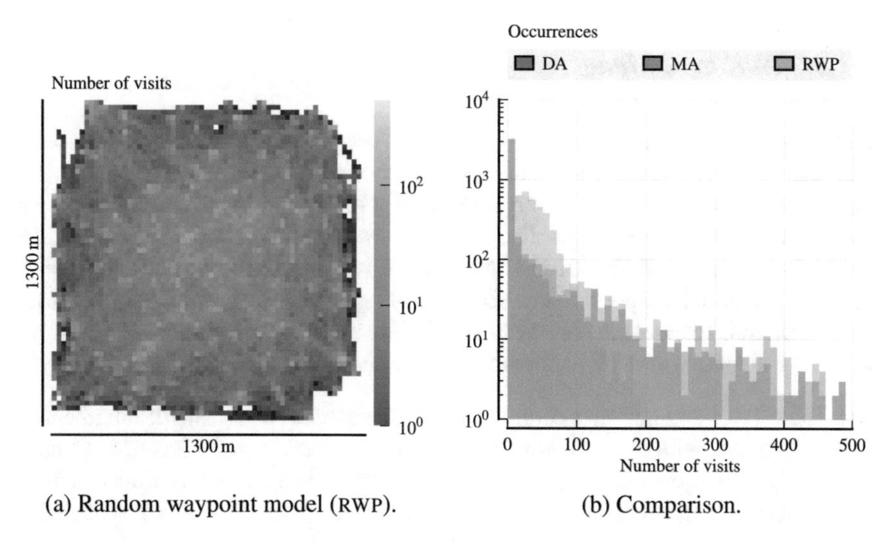

(a) Random waypoint model (RWP). (b) Comparison.

Fig. 6.3 Random waypoint mobility compared to map-based mobility

[3] www.pokemongomap.info [Accessed March 8th, 2017].

Table 6.2 Mobility parameters used in all models

Parameter	Default value
World size	$1,300\,\mathrm{m} \times 1,300\,\mathrm{m}$
Number of mobile clients	297
Movement speed $[v_{\min}, v_{\max}]$	$[1\,\mathrm{m/s}, 2\,\mathrm{m/s}]$
Pause times $\left[t_{\mathrm{p,min}}, t_{\mathrm{p,max}}\right]$	$[1\,\mathrm{s}, 60\,\mathrm{s}]$

with the Random Waypoint Model, as also pointed out in more recent research on human mobility models [9, 30].

All models are configured with the parameters listed in Table 6.2. This includes the size of the simulated area, the movement speed that is uniformly distributed between $v_{\min}$ and $v_{\max}$ for pedestrian mobility [14], and the time a client spends at the target location as previously explained. We select the size of the simulated area and the number of nodes according to the setup discussed in [28], motivated by a study by Kurkowski et al. [16] regarding the standardized evaluation of MANET routing protocols. However, as the respective studies are limited to random waypoint mobility, the respective characteristics in terms of node density and connectivity do not necessarily hold for our more complex mobility models. Therefore, we use the average number of subscribers resulting from our mobility models in combination with the workload models to characterize our simulation setup, as discussed in the following section.

6.1.2 Workload Models

Depending on the application at hand, interest in content can be closely tied to users' locations and their surrounding attraction points. In any case, users are interested in all events that are published to a location within a given distance from their current location. Within the publish/subscribe system, this interest is denoted via the `subscribe` method defined in Sect. 4.2.1, with the maximum distance being specified as the subscription radius. Each client periodically publishes an event to a target location. To assess the implications of different locality patterns, we define the following distinct workload patterns for a location-based mobile social application.

AR. The AR workload mimics communication characteristics of a mobile augmented reality game, with events being published to the user's current location. Consequently, only users in close proximity are interested in the event, leading to a high locality of content and interest. In this model, all events are published through the `publish_local` method as defined in Sect. 4.2.1.

LBS. The LBS workload models characteristics of a more generic location-based service. Clients publish events to specific attraction points rather than to their own location. Consequently, clients within proximity of the respective attraction

Table 6.3 Additional workload parameters

Parameter	Default value
Location update interval	5 s
Publication frequency per client	1/s
Application payload (added to events)	128 byte
Radius of the AoI	125 m

point are interested in the event. This model does not exhibit strong locality in content and interest. However, as events are being published to the locations of attraction points, the average number of interested subscribers is high. This is due to the fact that clients tend to gather around attraction points, as discussed in the previous section.

RND. As a generic baseline, we also consider a completely random choice of target locations. Consequently, some events might not have any subscribers at all, depending on the distribution of clients in the simulated area. For RND, all events are published through the `publish` method rather than through `publish_local`.

As discussed in [26], load on the communication system is directly determined by the mobility of clients in conjunction with the behavior of the selected workload model. Static configuration parameters used in all workload models are summarized in Table 6.3 and discussed in the following.

The location update interval describes the time between two consecutive location updates of a client's location sensor. This interval determines the maximum accuracy of location-based filter schemes, as discussed in Sect. A.1. The publication frequency is fixed to one publication per second per client. As the goal of this evaluation is the characterization of mechanism transitions and not a performance evaluation of individual mechanisms, we do not further vary the publication frequency. As individual mechanisms are already evaluated by their respective authors, we instead refer the interested reader to the respective publications.

The same holds true for the application payload added to outgoing events. Varying the payload has an influence on the performance of local dissemination mechanisms. When discussing the offloading effects resulting from the hybrid utilization of gateway selection and dissemination mechanisms, we compare the relative utilization of the respective mechanisms. Therefore, the concrete payload size does not have an impact on our reported results. Given that the frequency of events is fixed for each client, the load on the system increases significantly with increasing client density, as events have to be distributed to a potentially larger number of interested subscribers. With the workload models presented in this section, we assess the impact of these effects on a number of mechanisms within BYPASS.KOM.

As motivated previously, we use the average number of subscribers for individual publications to characterize our simulation setup. We show the respective spatial distribution for the DA and MA mobility model in Fig. 6.4, using the AR workload. The average number of subscribers varies significantly depending on the target location

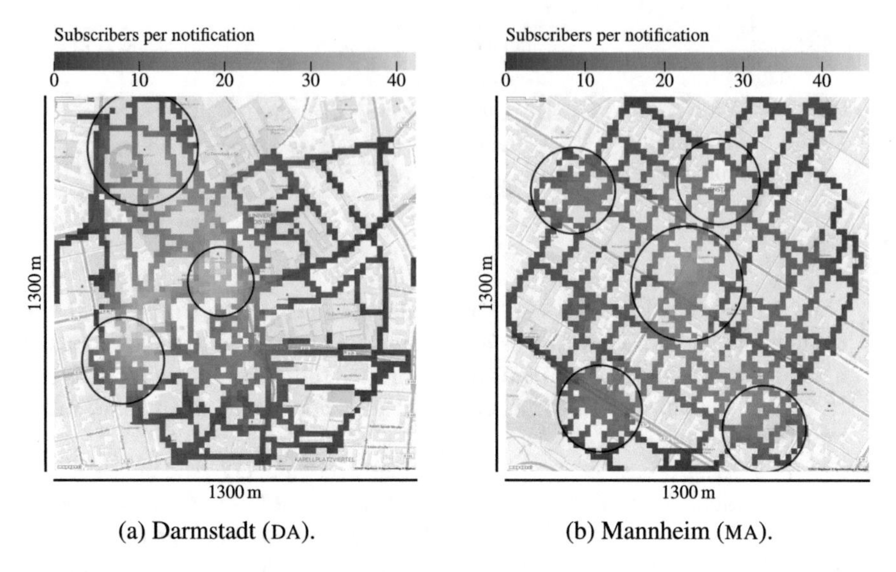

(a) Darmstadt (DA). (b) Mannheim (MA).

Fig. 6.4 Average number of subscribers based on the publication location

of a publication. Areas associated with attraction points (as indicated by the black circles) exhibit a higher client density and, consequently, higher number of interested subscribers. For both workloads, up to 40 clients within proximity are interested in a given publication, corresponding to approximately 13 % of the total client population. Rather than focusing on the absolute number of subscribers, we are more interested in the heterogeneity of this figure, as it motivates application-specific adaptation of the utilized mechanisms.

6.1.3 Network Model

To accurately model the scenario of mobile social applications, our simulation setup needs to provide two types of network models: (i) a model for direct ad hoc communication between mobile clients and (ii) a model for the communication with a cloud-based back end via a cellular connection.

For the direct ad hoc communication, we rely on the 802.11 Wi-Fi Medium Access Control (MAC) model with distributed coordination function as provided by the NS3 network simulator [27]. This model has been integrated into the PEERFACTSIM.KOM runtime within the SIMONSTRATOR.KOM platform as published in [20, 21]. The model relies on a simulation of the noise floor caused by concurrent transmissions to determine if packets can be transmitted. For unicast transmission, the respective RTS and CTS messages are exchanged before the actual data transmission occurs. For a more detailed description and analysis of the distributed coordination function of the 802.11 Wi-Fi MAC the interested reader is referred to [4]. The model relies on a log-distance path loss model configured with a path loss exponent of 3.8 for an urban environment as measured in the studies presented in [1, 17]. This configuration leads to an effective transmission range of around 88 m, while increasing the noise floor and, thus, the probability for collisions, for clients within up to 205 m range. As the chosen model takes the noise floor into account, it accurately captures the effect of increased client density and concurrent transmissions. Capturing this effect is crucial within the scope of our work, given that we utilize direct ad hoc communication at or around application-specific attraction points.

While an accurate model for the direct ad hoc connections between mobile clients is necessary to evaluate our work, we do not require the same level of accuracy for our model of the cellular network. Instead, we rely on a measurement-based approach for our model of the cellular network. A client's cellular connection to the cloud-based broker network is assumed to be reliable, which is a reasonable assumption in city centers as long as the mobile network is not overloaded. The model simply assigns an estimated bandwidth and latency to the respective connection based on real-world measurements presented and collected in [11, 12, 18, 19]. For technical details of the respective model, the interested reader is referred to our publication [26]. We assume that the resources of the cellular network are sufficient to cater for all mobile clients in our simulations. This is motivated by the fact that we are interested in the offloading effects that are achievable with our locality-aware event dissemination mechanisms rather than in the network behavior in overload situations. In our simulations, the cellular network has an average latency of 200 ms and offers reliable connectivity with a guaranteed bandwidth of 1 Mbps per client. We configured the bandwidth such that it does not impact the operation of our system (i. e., it ensures proper handling of the cellular traffic caused and consumed by mobile clients).

6.1.4 Evaluation Metrics

We utilize a number of metrics to capture the impact of mechanism transitions and the performance and cost characteristics of individual mechanisms. On the one hand, we capture performance and cost metrics specific to publish/subscribe systems, such as *recall* and *precision* of event delivery. On the other hand, we measure network load in terms of traffic for both, clients and the cloud entity, to assess the resource utilization of the system. Additionally, we capture mechanism-specific metrics to characterize the utilization of gateways where it is appropriate. Finally, to capture the effects of transitions on a system, we measure their execution time and location. In summary, the following metrics are gathered and reported in our evaluation:

Recall. The recall is defined as the ratio of correctly notified subscribers to all subscribers for a given event. Within our scenario, the valid set of subscribers for a given event is defined by the circular AoI provided by the application through our API. It is therefore independent of the actual filter scheme and should be close to 1.0 for all utilized mechanisms in order to satisfy the application requirements.

Precision. The precision complements the recall in characterizing a publish/subscribe system. It is defined as the ratio of correctly notified subscribers to all clients that were notified, regardless of whether they are actually subscribed to the event.

Delivery Delay. We measure the average time it took to forward an event from its publisher to all reached subscribers. The delivery delay strongly depends on how the network latency and bandwidth are configured, which is why the absolute delay is not as important as the relative increase or decrease for different dissemination mechanisms.

Traffic. Given the challenge to efficiently utilize resources of mobile clients and the cellular network, we measure the traffic characteristics of our system. We distinguish between traffic over the cellular interface and traffic caused by direct local ad hoc communication. Additionally, we measure the traffic caused by a filter scheme's context update protocol and the traffic resulting from the coordination of transitions.

Spread of Transitions. To understand the behavior of transitions, we capture the time of their local execution on each affected client together with the currently utilized mechanism. We further keep track of a client's current location when executing transitions, allowing us to characterize the spread of transitions within a scenario.

Gateway Utilization. We introduced local gateways in order to relieve the cellular network in Sect. 4.3.4. To assess the fairness characteristics of the respective selection algorithms, we measure how often and for how long clients are utilized as gateways and how many other clients are served by them.

To assess the behavior of transitions during runtime, metrics are reported over time. Two sample plots are shown in Fig. 6.5. In Fig. 6.5a, the mean of the metric (in this case, the recall) is plotted over time. The mean recall, plotted as a solid line, is calculated over all events that occur within ten seconds. Additionally, the shaded areas report the percentiles: the darker shaded area includes all values between the 25th and 75th percentile, and the light area includes values between the 5th and 95th percentile. The median (50th percentile) is included as a thin solid line for reference.

While this format is an accurate representation of value distributions over time, it is not well suited to compare a larger number of systems against each other. Therefore, systems included for comparison are often only plotted with their mean value, as shown in Fig. 6.5b. Both plots further include dashed vertical lines that mark the execution time of a transition affecting all clients in the system.

Whenever aggregated results are reported, box plots or bar plots are utilized, as shown in Fig. 6.6. The extent of the box indicates the 25th and 75th percentile, with the solid line indicating the median. Whiskers report the 2.5th and 97.5th percentile, meaning that 95 % of the observed values lie within the whiskers. Boxes report the data gathered during one run of an experiment, corresponding to one random seed. To report the behavior of the metric for the remaining repetitions, we plot a circle indicating the grand mean (mean of the mean) over all runs. Additional whiskers

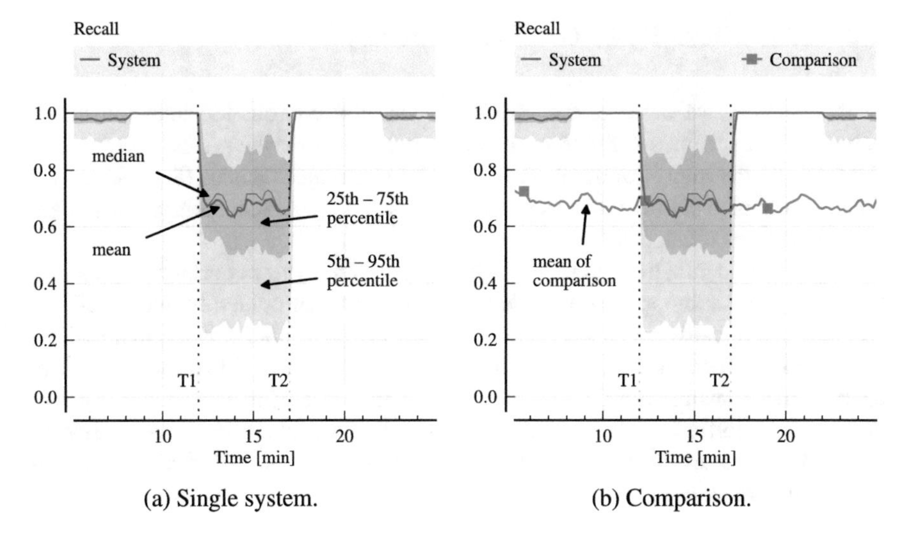

Fig. 6.5 Sample plots showing measurements over time

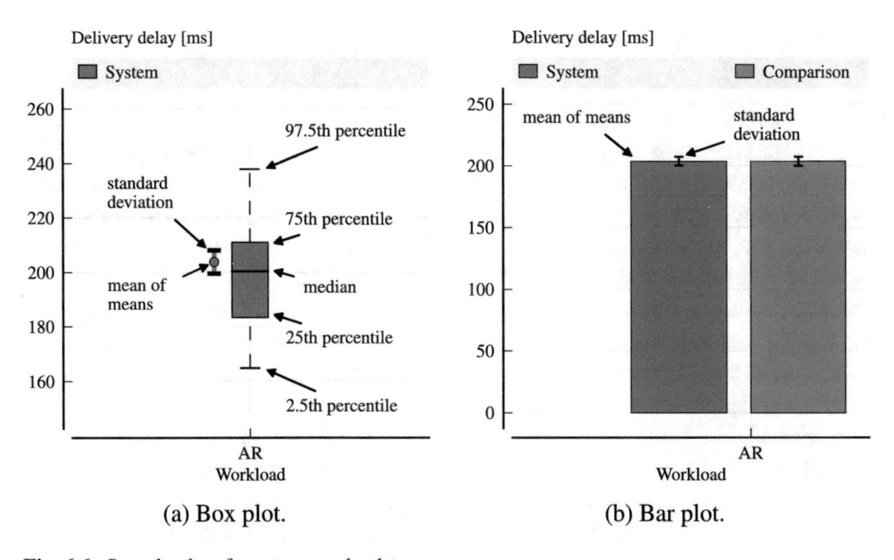

Fig. 6.6 Sample plots for aggregated values

around the circle indicate the standard deviation of the means for the individual runs of an experiment. Bar plots report the mean of the mean over multiple runs, with whiskers indicating the standard deviation of the individual mean values. Additional plot types are explained on their first occurrence in this chapter.

6.2 Transitions Between Filter Schemes

In this section, we evaluate the impact of centrally coordinated total transitions between filter schemes during runtime of the publish/subscribe system (Sect. 6.2.1). We focus on the impact of our proposed state transfer mechanisms (Sect. 6.2.2) and the potential of self-transitions as generic representation of scheme-specific parameter adaptations (Sect. 6.2.3).

In addition to total transitions, we study partial transitions between filter schemes based on application-specific knowledge about attraction points. In Sect. 6.2.4, we compare the transition-enabled system to a self-adaptive filter scheme that realizes the same filter behavior. This allows us to verify correct operation of our transition-enabled mechanisms and discuss the benefits and cost of mechanism transitions compared to self-adaptive mechanisms. An in-depth evaluation of the individual filter schemes realized within BYPASS.KOM and their parameter configurations is presented in Sect. A.1.

6.2.1 Centrally Coordinated Total Transitions

First, we evaluate total transitions between filter schemes as initially discussed in [24]. A total transition affects all clients, as defined in Sect. 4.2.3. We expect that the transition-enabled system utilizing a filter scheme A behaves like a static configuration of filter scheme A. After a transition to scheme B, we expect the overall system characteristics to reflect those of scheme B. We evaluate whether BYPASS.KOM meets these expectations and, thus, correctly executes transitions between filter schemes. The respective transitions are triggered at predefined times at the broker. The broker sends a corresponding execution plan to all currently subscribed clients. The transitions and the time of their execution are listed in Table 6.4.

Recall and precision for the transition-enabled system are shown in Fig. 6.7 for the AR workload. For comparison, static configurations of the respective filter schemes without transitions are also included in the plots. The recall does not vary much between the different schemes, except for the grid-based scheme GRID that has an average recall of 0.7. After switching to STE ($T1$) or to EGRID ($T3$), the recall reaches 1.0, while for RADIAL approximately 25 % of the clients experience a recall

Table 6.4 Execution plans for the evaluation of total transitions

#	Time (min)	Execution plan
$T1$	8	$[T : \text{RADIAL} \rightarrow \text{STE},\ T^{\bullet} : f(\text{STE}) = \alpha \leftarrow 1.5]$
$T2$	12	$[T : \text{STE} \rightarrow \text{GRID},\ T^{\bullet} : f(\text{GRID}) = \text{grid} \leftarrow 5]$
$T3$	17	$[T : \text{GRID} \rightarrow \text{EGRID},\ T^{\bullet} : f(\text{EGRID}) = \text{grid} \leftarrow 10]$
$T4$	22	$[T : \text{EGRID} \rightarrow \text{RADIAL}]$

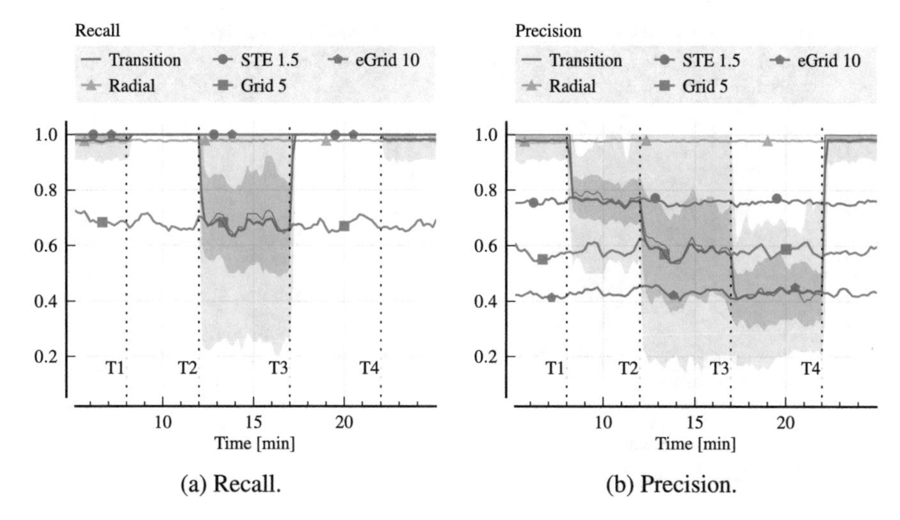

(a) Recall. (b) Precision.

Fig. 6.7 Recall and precision of the transition-enabled system over time

below 1.0, yet larger than 0.9. The precision of the transition-enabled system clearly reflects the characteristics of the individual filter schemes, as shown in Fig. 6.7b. As expected, the system adapts its characteristics to the respective filter scheme right after a transition occurs.

In Fig. 6.8, we show the traffic caused by the filter schemes' context update protocol. Traffic is shown over time to compare the performance of the transition-enabled scheme to the performance of the individual schemes after the execution of each total transition. Soon after each transition, the traffic caused by the context update protocol equals that of the currently active mechanism. However, we can observe sudden increases in traffic right after transitions $T2$ and $T3$ for the download direction and transitions $T3$ and $T4$ for the upload direction. This effect is caused by the bootstrap phase of the respective transitions. After transition $T2$ from STE to GRID the broker reports the assigned channel to each client, causing the sharp increase in context protocol download traffic for the client. For the same reason download traffic increases right after transition $T3$, as in EGRID the broker also sends assigned channels to clients. As each client is assigned to more than one channel in the EGRID scheme, the initial download is slightly larger due to increased message sizes.

Considering the upload direction shown in Fig. 6.8b, we observe an increase right after transition $T3$ and a slight increase right after transition $T4$. This is the consequence of missing state information at the broker, as discussed in detail in Sect. 6.2.2. The broker requires updated location information after $T3$ and $T4$ for channel assignment ($T3$) and general operation of RADIAL ($T4$). However, in both cases the previous filter scheme cannot provide the necessary information as neither GRID nor EGRID maintain updated client locations. Consequently, clients need to report their location if the assignment at the broker deviates too much from their current location, leading to the traffic increase right after $T3$. The effect is less

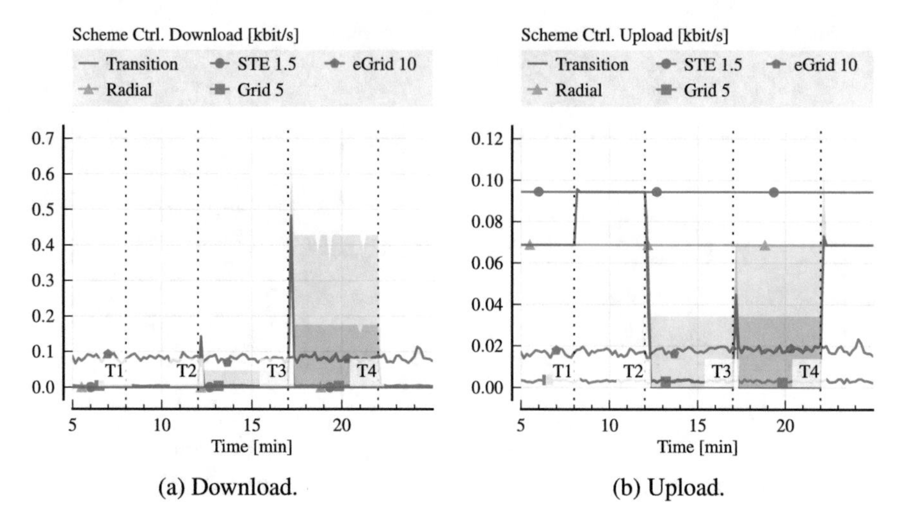

(a) Download. (b) Upload.

Fig. 6.8 Traffic characteristics of the transition-enabled system

visible for transition $T4$, as in RADIAL clients nevertheless need to send periodic location updates to the broker.

6.2.2 *Impact of the State Transfer Mechanism*

An important phase during the execution of a transition is the local state transfer between two mechanisms. As discussed in Sect. 4.2.3, state includes the subscriptions stored at the broker component of the respective filter scheme and any additional contextual information that has been gathered by the filter scheme. We compare the performance of the transition-enabled system with state transfer against transitions without state transfer for the AR and LBS workload models, assessing the difference between handling of local and generic location-based events during a transition. Without state transfer, clients need to re-subscribe at the broker and filter schemes need to be bootstrapped afterwards. Transitions are again triggered based on the elapsed time, with the execution plans given in Table 6.4.

Figure 6.9 shows the recall of both configurations over time right after each transition is triggered at the broker. Three cases can be distinguished and are discussed in the following: (i) the recall remains constant due to state transfer ($T1$ and $T2$), (ii) the recall drops to 0 for the AR workload even with state transfer in place ($T3$), or (iii) there is a slight quality degradation during the transition ($T4$). In the first case, shown in Fig. 6.9a, b the last known location of a client is information that is maintained by the original scheme's broker component *and* helps in bootstrapping the target scheme. After transition $T1$ from RADIAL to STE, the initial space time envelope is simply calculated with no additional movement vector, leading to a circular AoI. As

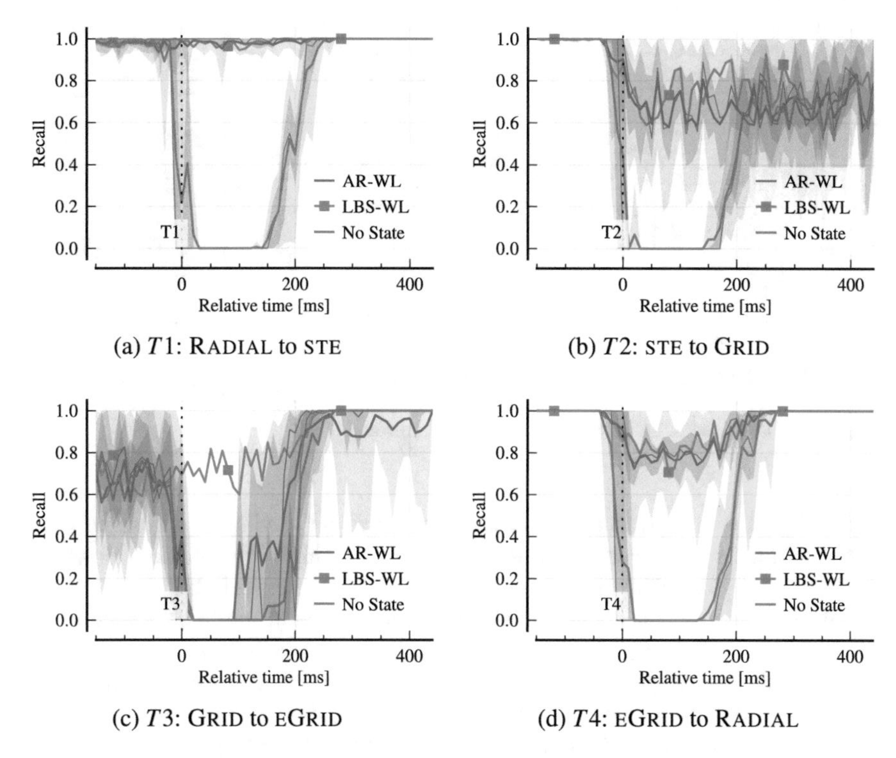

Fig. 6.9 Impact of state transfer with AR and LBS workloads during transitions

soon as clients report their movement vector, the broker updates the shape of the envelope. This, in turn, leads to an increase in achieved recall, as expected for the STE scheme (refer to Sect. A.1 for an in-depth evaluation of the individual schemes' properties). During the transition $T2$ from STE to GRID, information about the last known location of a client that is available within the STE filter scheme can directly be utilized for the channel assignment. Consequently, there is no intermediate drop in recall if state information is utilized, as shown in Fig. 6.9b.

After transition $T3$, the recall drops to 0 for a period of roughly 100 ms with state transfer, recovering afterwards. However, this is only the case for the AR workload publishing only local transitions. For generic location-based events, as created in the LBS workload, the state transfer mechanism is able to maintain the recall. This is due to the fact that the channel assignment changes from GRID to EGRID. Consequently, local events published to a specific channel in GRID can no longer be processed by the broker component of EGRID. For generic location-based events, assignment to channels is done by the broker. Given that the last known location of clients is transferred from GRID to EGRID, the broker can already use the updated channels of EGRID to notify relevant clients.

During transition $T4$, the slight quality degradation is caused by outdated state information being transferred to the RADIAL scheme. As EGRID only updates the last

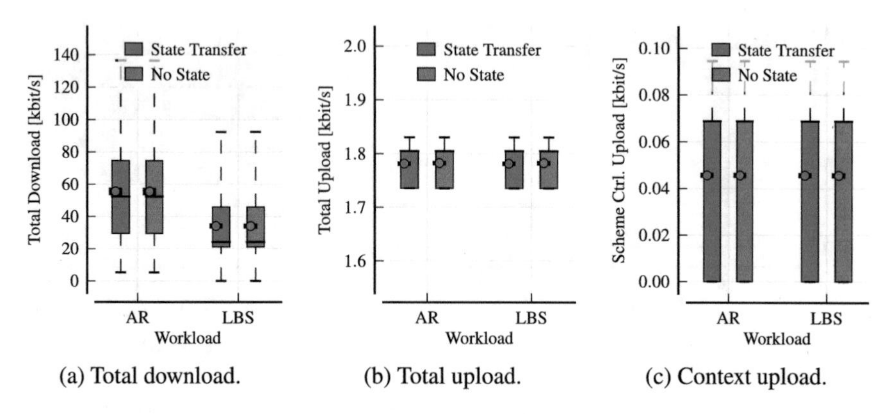

(a) Total download. (b) Total upload. (c) Context upload.

Fig. 6.10 Client-side traffic with and without state transfer

known location of a client as consequence of the client leaving its current cell, the location used as the center of the AoI in RADIAL is likely to be outdated. However, as soon as context update messages arrive at the broker, the recall reaches 1.0 again.

The state transfer mechanism does not affect the overall traffic compared to re-subscriptions, as shown in Fig. 6.10a, b. This is due to the fact that the traffic caused by context update messages and re-subscriptions is negligible compared to the traffic caused by client events and application payload within our scenario. The context update messages sent by clients account for only 2.5 % of the total upload, as shown in Fig. 6.10c. This value depends on the application payload included within events and on the frequency at which events are published. Given the asymmetry in up- and download traffic in our scenario, the download of context update messages accounts for even less overhead compared to the upload.

6.2.3 Self-transitions for Filter Scheme Adaptation

In addition to global transitions, we propose self-transitions as a generic design concept to adapt a mechanism's configuration. Within this section, we evaluate self-transitions of the grid-based filter scheme EGRID within the AR and LBS workload. Therefore, the execution plans as shown in Table 6.5 are triggered at the specified times. Initially, clients use the EGRID filter scheme with a grid factor of 2, leading to four cells. We discuss how updating the channel assignment function can lead to performance degradation during a transition for locally relevant events.

The resulting macroscopic system behavior under the AR workload compared to a static configuration of EGRID with the given grid factors is shown in Fig. 6.11. The transition-enabled system is able to adapt its characteristics to those of the individual configurations, as previously shown for total transitions. As expected for EGRID, the recall remains 1.0, until we execute a global transition to the RADIAL filter scheme

Table 6.5 Execution plans for the evaluation of self-transitions

#	Time (min)	Execution plan
$T1$	8	$[T^{\bullet} : f(\text{EGRID}) = \text{grid} \leftarrow 5]$
$T2$	12	$[T^{\bullet} : f(\text{EGRID}) = \text{grid} \leftarrow 10]$
$T3$	17	$[T^{\bullet} : f(\text{EGRID}) = \text{grid} \leftarrow 20]$
$T4$	22	$[T : \text{EGRID} \rightarrow \text{RADIAL}]$

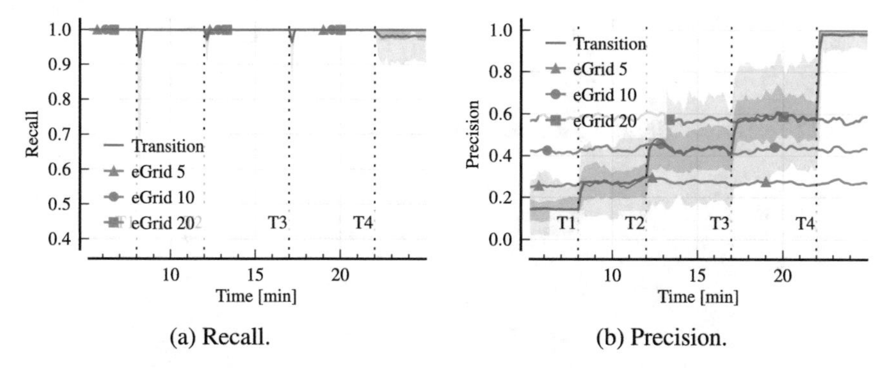

(a) Recall. (b) Precision.

Fig. 6.11 Recall and precision with self-transitions compared to static configurations

($T4$), as shown in Fig. 6.11a. However, there are slight quality degradations right after transitions are executed, clearly visible for $T1$ and $T3$. The precision follows the exact characteristics of static configurations as shown in Fig. 6.11b, leading to a stepwise increase for smaller cells caused by higher values of the grid factor.

To identify the reasons for the quality degradation right after each transition, we take a closer look at $T1$ and $T3$ under both workloads, AR and LBS. From Fig. 6.12 it becomes clearly visible that the performance is only degraded for the AR workload, whereas the performance for the LBS workload is not affected. This is due to the fact that events published to the client's own location (as issued within the AR workload) are published to the channel where the client is currently located. This ensures that all subscribers registered for that channel receive the event, as discussed in Sect. 4.2.2. However, as a consequence of the self-transition, the channel assignment is updated. This, in turn, leads to new channel identifiers being issued by the broker. Until clients receive the updated channel identifiers via the context update protocol, incoming events using the outdated channels are simply discarded. The duration of this effect is determined by the latency of the cellular connection, which is configured to around 200 ms, as discussed in Sect. 6.1.3.

For non-local events as issued by the LBS workload, the assignment to a target channel is not performed by the client, but instead by the broker. Therefore, incoming events can directly be processed with the updated assignment function and forwarded to the correct subscribers. As discussed in Sect. 4.2.3 for the case of missing context information at the broker, this effect can be mitigated by maintaining old channel

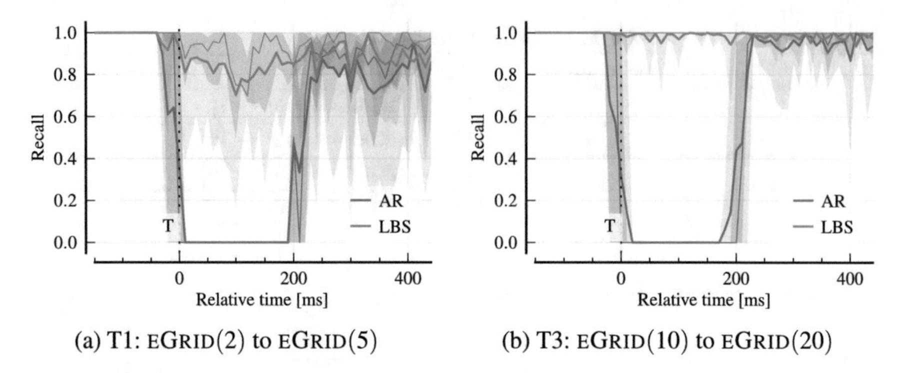

Fig. 6.12 Recall of EGRID during self-transitions $T1$ and $T3$

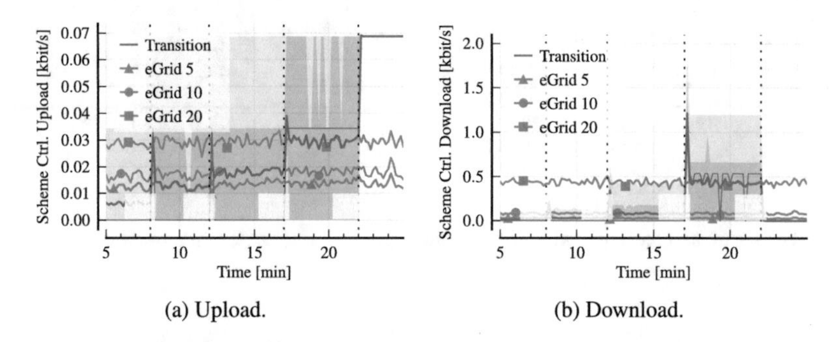

Fig. 6.13 Client-side control traffic caused by self-transitions

assignments in addition to the new ones for some time after a transition occurred. Such a mechanism would already be part of a more sophisticated self-adaptive version of EGRID.

As shown in this evaluation for the case of the channel-based scheme EGRID, self-transitions allow us to benefit from self-adaptability of filter schemes. Given the generic nature of self-transitions, we can easily integrate the potential of self-adaptations into an existing transition execution plan. This enables even more precise control over the performance and cost characteristics of the transition-enabled publish/subscribe system (Fig. 6.13).

6.2.4 Partial Transitions versus Self-adapting Mechanisms

By utilizing simple mechanisms and executing transitions between these mechanisms, we aim to achieve high adaptivity under dynamic environmental conditions. In the following evaluation scenario, we compare the performance and cost of filter scheme transitions to a self-adaptive filter scheme that offers the same functionality.

Transitions are executed between the parametric scheme RADIAL and the channel-based scheme ATTRACT whenever a client approaches an attraction point.

The scenario is described in detail in Sect. 4.4 for coexisting transition-enabled mechanisms. However, in contrast to the execution plans discussed there, we only switch between the RADIAL and ATTRACT filter scheme. We neither use a local dissemination mechanism nor a gateway selection algorithm. Consequently, as soon as a client u approaches an attraction point and enters the area covered by a channel in ATTRACT, the partial transition $T^{\parallel}(u)$: RADIAL $\rightarrow$ ATTRACT is executed by the broker. Additionally, the client executes the total transition T : RADIAL $\rightarrow$ ATTRACT. Accordingly, T : ATTRACT $\rightarrow$ RADIAL is executed by the client when leaving the area covered by the channel. In both cases, available state is transferred in-between the broker components of both filter schemes during the partial transition.

The behavior of the transition-enabled system is compared to that of MULTI, a filter scheme that realizes the same behavior within one mechanism. MULTI is an extended version of ATTRACT, where clients that are not assigned to one of the attraction points report their current location to the broker. The broker filters incoming events based on the reported location whenever a client is not assigned to a channel.

Figure 6.14 shows filter scheme transitions based on the client's current location when executing a transition. The radii of the attraction points within ATTRACT are shown as black circles for reference. Each colored dish corresponds to a transition to the respective mechanism identified by the color. Transitions are aggregated for cells of 10×10 m, with the radius of the circle indicating the number of transitions within the cell. As expected, transitions between RADIAL and ATTRACT are executed when clients enter or leave the area around the respective attraction point.

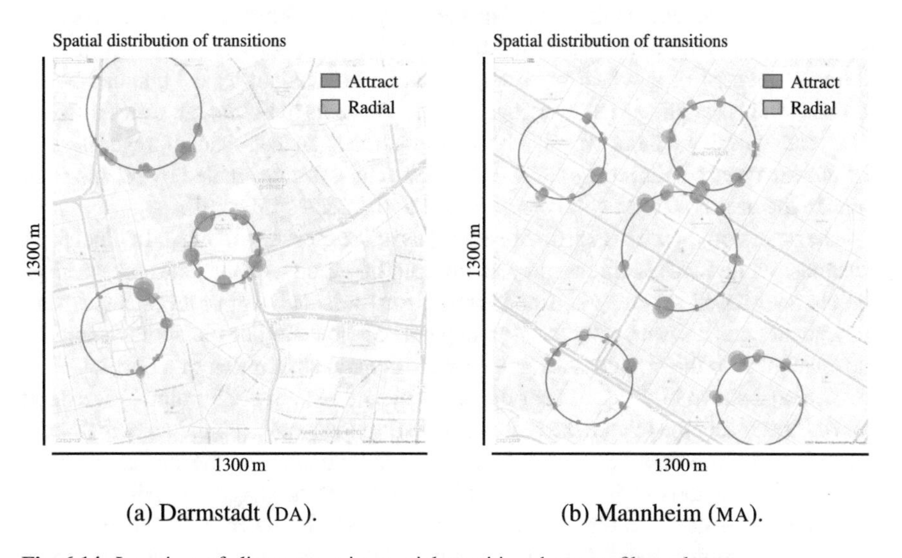

(a) Darmstadt (DA). (b) Mannheim (MA).

Fig. 6.14 Locations of clients executing partial transitions between filter schemes

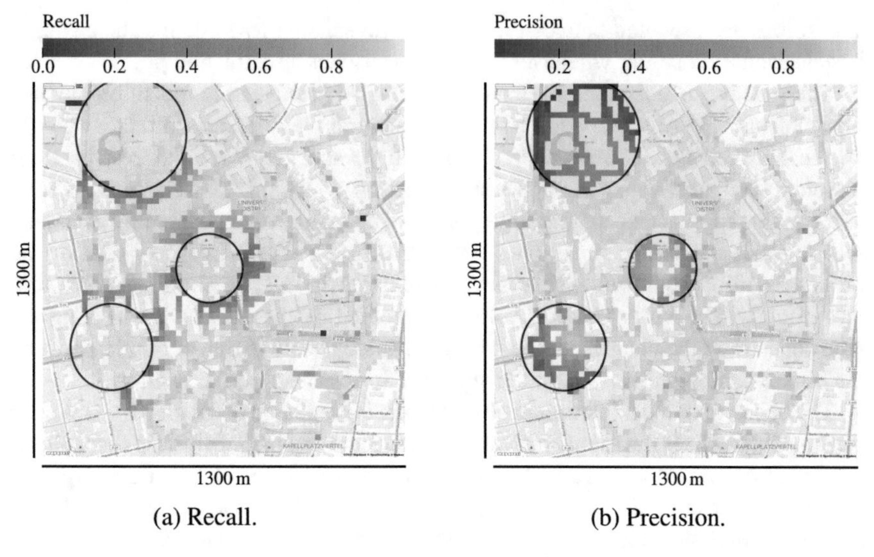

(a) Recall. (b) Precision.

Fig. 6.15 Performance of the transition-enabled system for the AR workload

The resulting performance characteristics for the transition-enabled system depending on the target location of an event are shown in Fig. 6.15 for the Darmstadt scenario. The recall within the area covered by channels in ATTRACT (indicated by the black circles) is approximately equal to 1.0. However, in direct vicinity of the respective area around an attraction point, the recall within the RADIAL scheme degrades significantly as shown in Fig. 6.15a. Further away from any attraction points, RADIAL is again able to maintain a recall of 1.0. This behavior is caused by the fact that the radius of interest of a subscription is not taken into account by the channel-based scheme when assigning incoming location-based events to a channel. Consequently, all clients within a channel receive only events issued to that channel, regardless of the area covered by their original subscription. Thus, events created by clients right outside the area covered by the channel lead to the decreased recall.

The precision shown in Fig. 6.15b reflects the characteristics of the individual filter schemes: while RADIAL allows precise filtering based on a client's current location, the channel-based scheme ATTRACT simply forwards an event to all clients within the channel. As a result, the precision depends on how similar the area covered by the channel is to the circular AoI given by the actual subscription of a client.

To also achieve high recall near the area covered by attraction points, we extend the ATTRACT scheme to take the radius of subscriptions into account when matching incoming location-based events. This is achieved by increasing the spatial extent of the channel with the maximum radius of interest out of the stored subscriptions. The resulting performance in terms of recall and precision is shown in Fig. 6.16. We no longer observe lower recall around areas covered by a channel in ATTRACT, as shown in Fig. 6.16a. Instead, the recall remains constantly high for all areas on the map. This is achieved at the cost of lower precision, as shown in Fig. 6.16b. Events that are

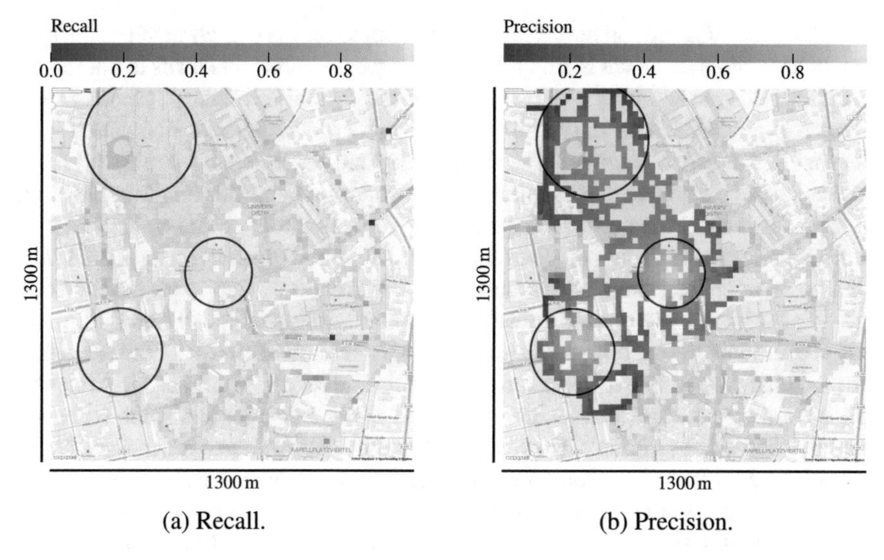

(a) Recall. (b) Precision.

Fig. 6.16 Extending channel coverage by taking the radius of subscriptions into account

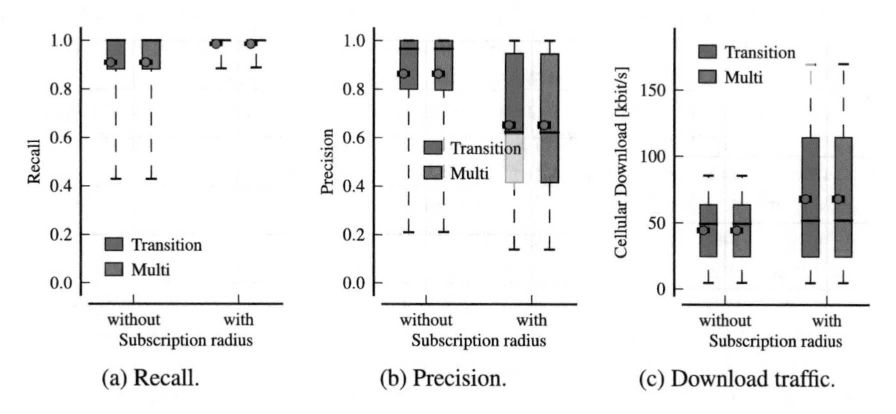

(a) Recall. (b) Precision. (c) Download traffic.

Fig. 6.17 Performance and cost comparison

published within the additional area defined by the radius of interest of subscriptions are now forwarded to all subscribers of the channel as well.

For easier comparison, we show the aggregated performance and cost of both variants in Fig. 6.17. As the effect does not only affect the transition-enabled system, but also the adaptive filter scheme MULTI, we report the respective results for both filter schemes with and without taking the radius of interest into account. As expected, both systems exhibit the exact same characteristics, showing that the functionality of a self-adaptive filter scheme can be realized using partial transitions between two less complex filter schemes. Additionally, this comes at virtually no additional costs in terms of traffic or decreased performance.

Taking the subscription radius into account improves the overall recall of the system in both cases as shown in Fig. 6.17. The decrease in precision leads to additional cellular traffic caused by the transmission of unnecessary events for clients in the respective channels. This only affects clients that are within the area covered by a channel in ATTRACT, leading to the skewed distribution observable in Fig. 6.17c.

The radius associated to a location-based subscription depends on the application at hand. This radius is constant for all mobile clients in our scenario, leading to a rather simple determination of the extension area required to achieve a recall of 1.0. If equal radii of interest cannot be assumed, one has to choose an appropriate extension area that achieves the desired trade-off between achievable recall and additional traffic caused by the decrease in precision. To guarantee a recall of 1.0, the extension of the channel area has to match the maximum radius out of all subscriptions by clients within the respective channel. One can easily calculate this value based on the set of active subscriptions. However, each client entering or leaving the area could potentially alter the current maximum value, requiring frequent recalculation. Given the application-specific nature of attraction points, we believe that it is reasonable to assume that a suitable radius for such extension areas could be defined by the application beforehand.

We only reported results for the Darmstadt scenario in this section, as they match the behavior of the system in other conditions. For completeness, we report additional results for the Mannheim scenario and other workload models in A.2.

In this section of our evaluation we demonstrate the applicability of mechanism transitions to location-based filter schemes for publish/subscribe systems. We evaluate three distinct types of transitions as introduced in Chap. 4: (i) *total transitions* affecting all clients of a broker, (ii) *self-transitions* used for generic adaptation of a mechanism's state, and (iii) *partial transitions* that allow parallel operation of multiple filter schemes for different groups of clients. Transitions between filter schemes enable the publish/subscribe system to adapt the performance versus cost trade-off depending on the current application requirements or environmental conditions. We specifically study the impact of our proposed state transfer mechanism on the execution of transitions, showing its significance in achieving seamless transitions without degradations in performance. Partial transitions enable the utilization of application-specific knowledge by switching filter schemes, for example, for clients that approach an attraction point. Combined with the evaluation of transition-enabled dissemination mechanisms in the following section, these insights form the basis for the evaluation of coexisting mechanisms presented in Sect. 6.4.

6.3 Transitions Between Event Dissemination Mechanisms

In addition to filter schemes for location-based filtering, BYPASS.KOM includes mechanisms for locality-aware event dissemination. Depending on their configuration and the execution of transitions, BYPASS.KOM can be used in a pure ad hoc fashion or in a hybrid mode that relies on a cloud-based broker. In the following section, we assess achievable savings in terms of offloading the cellular connection by performing selected aspects of the brokering process directly on mobile clients. Therefore, we evaluate individual ad hoc dissemination mechanisms and different modes of hybrid operation as proposed in Sect. 4.3.2.

In Sect. 6.3.3 we conduct a proof-of-concept evaluation of centrally coordinated transitions between dissemination mechanisms. The central coordination of transitions between local dissemination mechanisms might lead to unnecessary overhead, given that clients already communicate locally. Therefore, in Sect. 6.3.4, we evaluate the utilization of the self-healing mechanism presented in Sect. 4.3.3 to spread transitions within an ad hoc network. Understanding both concepts of coordinating transitions between dissemination mechanisms lays the foundation for the evaluation of coexisting transition-enabled mechanisms in Sect. 6.4.

6.3.1 Pure Ad Hoc Event Dissemination

Before evaluating the performance of the proposed hybrid dissemination modes, we first characterize the individual ad hoc dissemination mechanisms in BYPASS.KOM. Their performance is evaluated within the Darmstadt scenario, and aggregated results for the AR and RND workloads are shown in Fig. 6.18. While RND publishes events to arbitrary locations, the AR workload only issues local events, allowing the dissemination mechanism to utilize geofencing to limit the spread of messages. Additionally, events in the AR workload only need to reach clients within vicinity of the publisher to achieve a high recall, while in RND they need to reach the respective random target location. This location might only be reachable via multiple intermediate hops or it might not be reachable at all if network partitions occur. In addition, as messages are not subject to geofencing, the overall load on the network increases significantly for RND compared to the AR workload.

The results reported in Fig. 6.18 support this rough assessment regardless of the respective dissemination mechanism. In general, the average achievable recall for pure ad hoc dissemination in the RND workload lies well below 0.4, with GOSSIP and HYPERG achieving an average recall of below 0.2, as reported in Fig. 6.18a. The single-hop dissemination mechanism BROADCAST is obviously not suited to disseminate events to arbitrary target locations, with an average of 1 % of all events reaching their intended receivers. The more aggressive schemes FLOODING and PLAN-B perform best for random target locations, yet reaching only approximately 35 % of all subscribers on average.

If events are only to be disseminated to nearby mobile clients (as is the case in the AR workload) all multi-hop ad hoc dissemination mechanisms perform reasonably well. While GOSSIP and HYPERG achieve an average recall of 0.85 and 0.8, respectively, with FLOODING and PLAN-B approximately 95 % of all subscribers are notified. The precision shown in Fig. 6.18b illustrates the impact of geofencing within the AR workload. As events in the RND workload essentially need to be spread to every mobile client, the resulting precision is expected to be low, even in the face

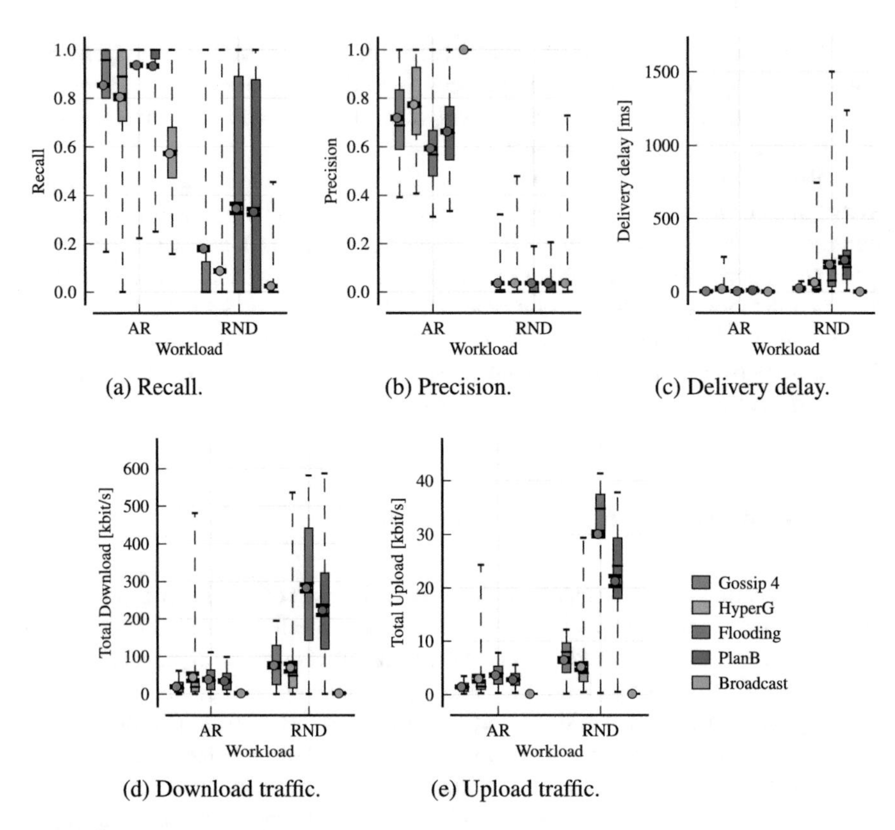

Fig. 6.18 Performance and cost of pure ad hoc dissemination

of a generally low recall. With geofencing, only clients that are actually subscribed to an event that is being disseminated forward it, leading to an average precision of between 0.6 and 0.8 for the multi-hop dissemination mechanisms. The high precision of the single-hop mechanism BROADCAST for the AR workload is expected, given that events only reach clients within direct vicinity of the producer.

The delivery delay of ad hoc dissemination mechanisms shown in Fig. 6.18c is generally in the order of a few milliseconds, provided that the network is not overloaded. However, once the network is overloaded (as is the case in the RND workload for HYPERG, FLOODING, and PLAN-B) the delay increases significantly. At the same time, the probability of collisions during the wireless transmission increases, leading to the aforementioned drop in achievable recall. Figure 6.18d shows the download traffic per client for both workloads. The effect of geofencing is clearly visible for the multi-hop dissemination mechanisms. On average, the download traffic of mobile clients increases by a factor of six if the spread of messages is no longer limited by geofencing. The same effect can also be observed for the average upload traffic for each mobile client. Here, the aggressive forwarding policies of FLOODING

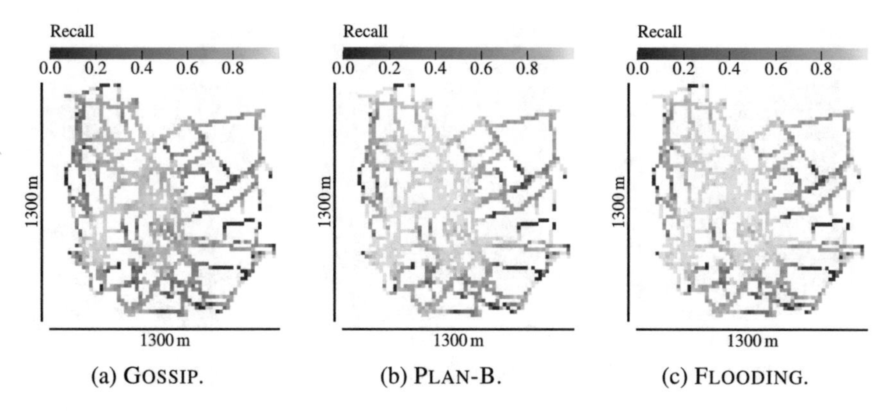

(a) GOSSIP. (b) PLAN-B. (c) FLOODING.

Fig. 6.19 Recall depending on the target location of an event, AR workload

and PLAN-B result in a significant increase in upload traffic compared to the less aggressive mechanisms GOSSIP and HYPERG.

The performance of an ad hoc dissemination mechanism depends on the availability of sufficient forwarders within proximity. Therefore, we also study the recall as a location-dependent metric for the individual dissemination mechanisms. The results for the AR workload are shown in Fig. 6.19. Plots show the average recall for events based on their target location, resulting in measurements at pedestrian walkways for the AR workload.

All mechanisms perform better in areas with higher client density compared to more remote areas with lower client density on the map.[4] Also, as discussed previously, PLAN-B and FLOODING achieve a higher recall due to their more aggressive transmission strategy compared to GOSSIP. However, even aggressive mechanisms are unable to maintain their high recall at more remote areas of the map. This is even more significant within the RND workload, shown in Fig. 6.20. Here, we cannot benefit from geofencing when disseminating events, leading to a general increase in the number of messages and, thus, collisions in the network, as previously discussed. Still, the effect of client density on the achievable recall is clearly visible for all dissemination mechanisms.

For comparison, we also report the results for the RND workload with the random waypoint mobility model in Fig. 6.21. The average client density is too low for GOSSIP with a forwarding probability of $p = 0.4$, leading to a generally low recall as shown in Fig. 6.21a. FLOODING and PLAN-B benefit from their aggressive transmission strategy and the rather homogeneous distribution of clients, achieving high recall in the central region of the simulated area.

The results presented in this section motivate the utilization of ad hoc dissemination mechanisms for transmission over a few hops and in rather dense scenarios. They also clearly show that pure ad hoc dissemination of events is not suitable, especially

[4] For a discussion of the client density resulting from the Darmstadt mobility model please refer to Sect. 6.1.1.

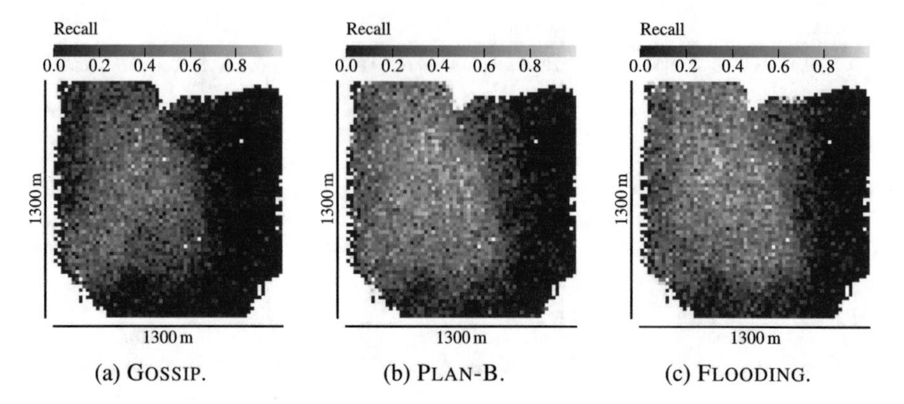

(a) GOSSIP. (b) PLAN-B. (c) FLOODING.

Fig. 6.20 Recall depending on the target location of an event, RND workload

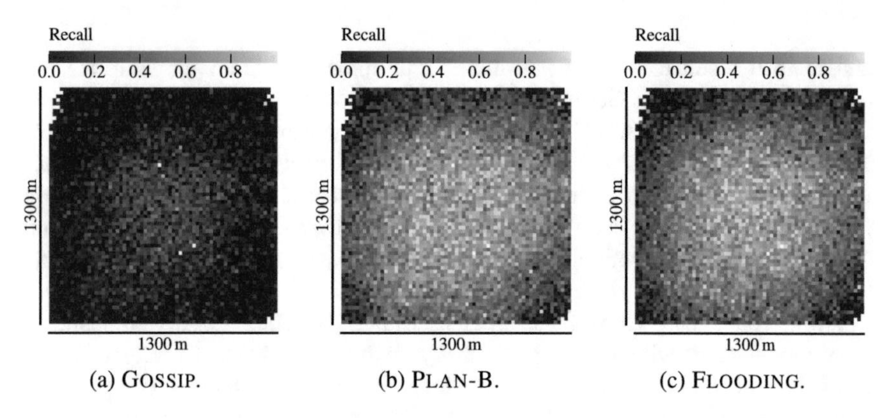

(a) GOSSIP. (b) PLAN-B. (c) FLOODING.

Fig. 6.21 Recall achieved under the random waypoint mobility model, RND workload

if events are not only relevant near their producer. Within BYPASS.KOM, we therefore consider two distinct application areas for ad hoc dissemination of events: (i) *hybrid operation modes* as evaluated in the following section, and (ii) *dissemination around application-specific attraction points* as evaluated in Sect. 6.4.

6.3.2 Hybrid Dissemination Mechanisms

In this section, we discuss the cost and benefits of the hybrid dissemination modes introduced in Sect. 4.3.2 and [20] under different workload models. Understanding the characteristics of hybrid event dissemination serves as the foundation for the following discussion of transitions between event dissemination mechanisms, especially with respect to their coordination.

Table 6.6 Hybrid dissemination modes studied in the evaluation

Mode	Ad Hoc dissemination	Cloud dissemination	Gateways
OFFLOAD	Local	Non-local	No
AUGMENT	Local	Local and Non-local	No
GATEWAY	Local	Local and Non-local	Yes
AD HOC	Local and Non-local	None	No
CLOUD	None	Local and Non-local	No

The modes and their key differences with respect to the dissemination of events are summarized in Table 6.6. The three modes OFFLOAD, AUGMENT, and GATEWAY utilize direct ad hoc dissemination for local events. In our evaluations, we rely on the GOSSIP dissemination mechanism with geofencing enabled as described in Sect. 4.3.1. While in OFFLOAD only non-local events are sent to the cloud for further dissemination, the cloud is used for all events in the remaining two hybrid modes. When sending events to a group of clients, GATEWAY utilizes local gateways that notify their one-hop neighbors via broadcast (BROADCAST). For the GATEWAY mode, we utilize SKC_LEACH as gateway selection algorithm. AD HOC and CLOUD are included for reference, given that they represent the two extreme cases of purely local and purely cloud-based event dissemination.

Figure 6.22 reports the performance metrics for the aforementioned modes within the Darmstadt mobility model. To assess the impact of local and non-local events, we report the results for the AR and the LBS model, as introduced in Sect. 6.1.2. Considering the recall of 0.2 for AD HOC as reported in Fig. 6.22a, pure ad hoc dissemination of events is not suitable for the generic location-based workload LBS. The potentially large distance between producer and consumer of an event leads to an increased probability of intermediate message loss. Additionally, the MANET is not guaranteed to be fully connected at all times, as partitions might occur due to client mobility. Even if events are to be distributed solely to clients within vicinity (as is the case in the AR workload) the recall reaches only 0.85. Utilizing a more sophisticated (or aggressive) local dissemination mechanism can increase the recall in this case. However, this comes at the cost of increased communication overhead and traffic in the MANET, potentially increasing the number of collisions.

The precision reported in Fig. 6.22b supports the claim that a pure ad hoc dissemination is not suitable for our scenario. The geofencing mechanism for local events limits the spread of messages to the vicinity of the interested clients, thus leading to an average precision of 0.7 in the AR workload. For generic location-based events as issued in the LBS workload, a larger number of clients is notified on the path towards the intended group of receivers. This leads to the significant decrease in precision for the AD HOC dissemination mechanism.

Compared to a pure cloud-based dissemination mechanism, the AD HOC mode substantially speeds up the delivery of events. This effect is best visible if events are to be distributed only in the local neighborhood, leading to only a few milliseconds

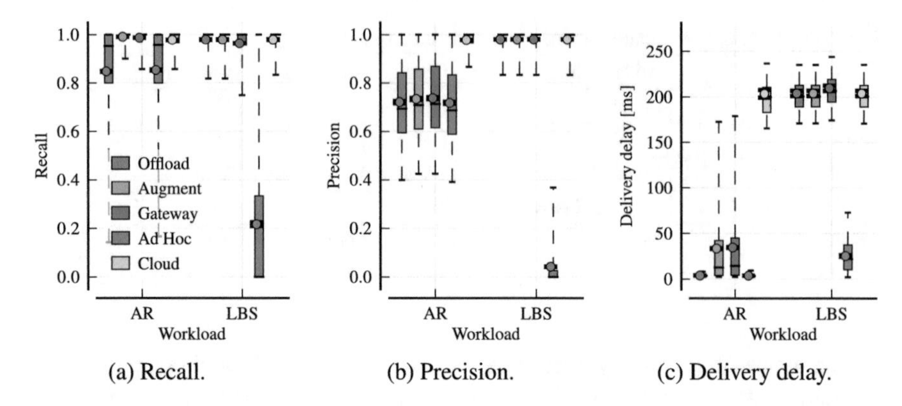

(a) Recall. (b) Precision. (c) Delivery delay.

Fig. 6.22 Performance of hybrid event dissemination mechanisms

of average dissemination delay as reported in Fig. 6.22c. In comparison, a pure cloud-based dissemination takes an average of 200 ms caused by the network delay between mobile user and cloud-based back end. Given a reliable cellular connection, recall and precision for CLOUD are nearly optimal for all workloads. The slight degradations in recall and precision for the cloud-based dissemination mode are caused by outdated state information of the underlying filter scheme. Clients report their current location only once every five seconds with the default configuration of the RADIAL filter scheme used in this evaluation. Therefore, events that are published right at the edges of a client's AoI might not yet (or no longer) be covered by the AoI calculated at the broker based on the last reported location.

In the following, we discuss the hybrid modes of operation compared to the CLOUD and AD HOC operation mode. Here, the AD HOC mode serves as the lower bound in terms of achievable delivery delay, and the CLOUD mode as an upper bound in terms of recall and precision. As expected, the OFFLOAD mode behaves as AD HOC in case there are only local events, as these events are not sent to the cloud. Consequently, it behaves like CLOUD for the LBS workload, as all non-local events are sent to the cloud without any ad hoc dissemination being utilized.

The AUGMENT mode achieves an even higher recall than CLOUD in case of the AR workload. This is due to the fact that local dissemination is not affected by the previously discussed problem of outdated state information at the broker. Instead, affected clients still receive the event via the ad hoc dissemination mechanism in most cases, slightly increasing the recall. Given that not only interested, but also some uninterested clients receive the locally disseminated event, the precision of AUGMENT for the AR workload resembles that of the pure ad hoc mode. For the LBS workload, we cannot distinguish between AUGMENT and OFFLOAD, as both modes differ only in how they handle local events.

When utilizing gateways to distribute events to a group of local clients, there is a slight decrease in the observed recall compared to the AUGMENT mode. This, again, is caused by outdated state information during calculation of a suitable set

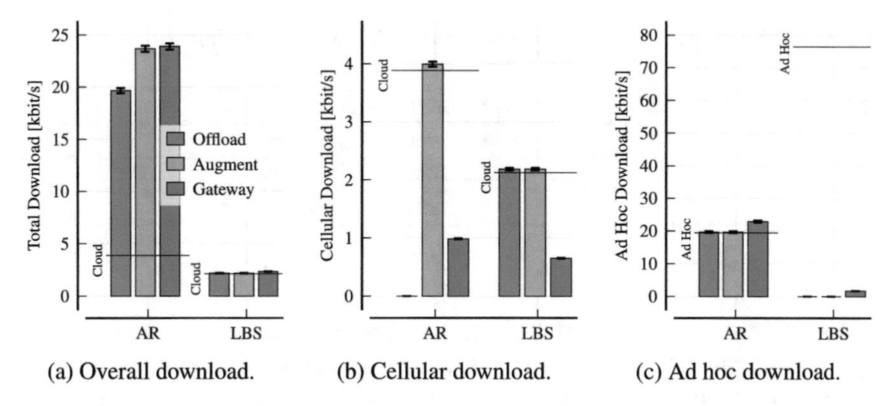

(a) Overall download. (b) Cellular download. (c) Ad hoc download.

Fig. 6.23 Per-client download traffic caused by hybrid dissemination mechanisms

of gateways at the broker, as the selected gateways might not be able to reach all interested clients. We further explore this effect in Sect. 6.4 for coexisting transition-enabled mechanisms. Although the GATEWAY mode does not affect the performance metrics reported in Fig. 6.22, it has a significant impact on the traffic characteristics.

Figure 6.23 shows the traffic caused for clients by each of the hybrid modes, compared to the respective baseline. Considering the cellular download traffic as shown in Fig. 6.23b, the pure cloud-based dissemination serves as the upper bound in cellular traffic, as indicated by the solid lines showing the average value.[5] As expected, OFFLOAD and AUGMENT follow the traffic characteristics of CLOUD and AD HOC depending on whether local or non-local events are disseminated. With the utilization of gateways in the hybrid mode GATEWAY, the cellular download traffic per client is reduced to about 25 % compared to CLOUD. As discussed previously, this reduction in cellular download traffic has no significant impact on the achieved recall for both workloads. At the same time, the addition of gateways only causes a slight increase in the ad hoc download traffic, as shown in Fig. 6.23c.

Utilizing gateways in the event dissemination process contributes to the goal of locality-aware dissemination, as the gateways essentially act as local brokers for their direct vicinity. Their utilization avoids redundant transmissions of events to a potentially large number of clients within close proximity. Consequently, gateways constitute a way to utilize locality of interest within the process of event brokering. We analyze the impact of gateways in more detail in Sect. 6.4.2 in the context of coexisting mechanisms. There, we also discuss the fairness characteristics of different gateway selection algorithms utilized within our work.

[5] Actually, the hybrid configurations exhibit a slightly higher cellular download traffic than the pure cloud-based dissemination method. This is caused by additional information included in the message header. In all hybrid schemes, clients report two IP addresses to the cloud: one used by the cellular interface and the other used for local ad hoc communication. This additional information slightly increases the resulting message size. The effect is only visible due to the fact that the application payload is rather small, as defined in Sect. 6.1.2.

Table 6.7 Execution plans for transitions between dissemination mechanisms

#	Time (min)	Execution plan
$T1$	8	$[T$: GOSSIP → PLAN-B$]$
$T2$	12	$[T$: PLAN-B → HYPERG$]$
$T3$	17	$[T$: HYPERG → FLOODING$]$
$T4$	22	$[T$: FLOODING → BROADCAST$]$

Based on our evaluation of individual ad hoc dissemination mechanisms and hybrid dissemination modes, we evaluate the execution of transitions between different dissemination mechanisms in the following section. Executing transitions between dissemination mechanisms enables us to later combine the selection of suitable gateways with ad hoc dissemination mechanisms in a locality-aware fashion.

6.3.3 Globally Coordinated Transitions

Given the cloud entity in any of the hybrid dissemination modes, our first approach to transitions between dissemination schemes relies on central coordination. As discussed previously, this is realized by disseminating transition execution plans to all affected clients via the cellular network. We focus on the execution of transitions and its impact on the overall system characteristics, addressing the last step of the MAPE-cycle introduced in Sect. 2.4. The analysis and planning steps are not within the scope of our work. To study the execution of transitions, we therefore trigger transitions at predefined times, as summarized in Table 6.7.

Figure 6.24 shows the resulting recall and precision of the transition-enabled system over time for the AR workload. For comparison, the average values for each individual mechanism are included. As for the case of transition-enabled filter schemes discussed previously, the transition-enabled system adapts its characteristics to the individual mechanism after each transition rapidly. As these characteristics are discussed in detail in Sect. 6.3.1, we instead focus on the timing of transitions and the system's behavior during their execution in the following. Given the bad performance of pure ad hoc event dissemination without geofencing, we only discuss the transition-enabled system under the AR workload in this section. For a discussion of the performance and cost of local dissemination mechanisms under other workload and mobility models, please refer to Sect. A.3.

In contrast to the case of filter schemes, we do not transfer any state information between individual dissemination mechanisms. Additionally, incoming messages can only be processed by a client if they were created with the same dissemination mechanism, potentially leading to message loss during a transition. As shown in Fig. 6.25b, all clients within the given scenario execute the transitions within a

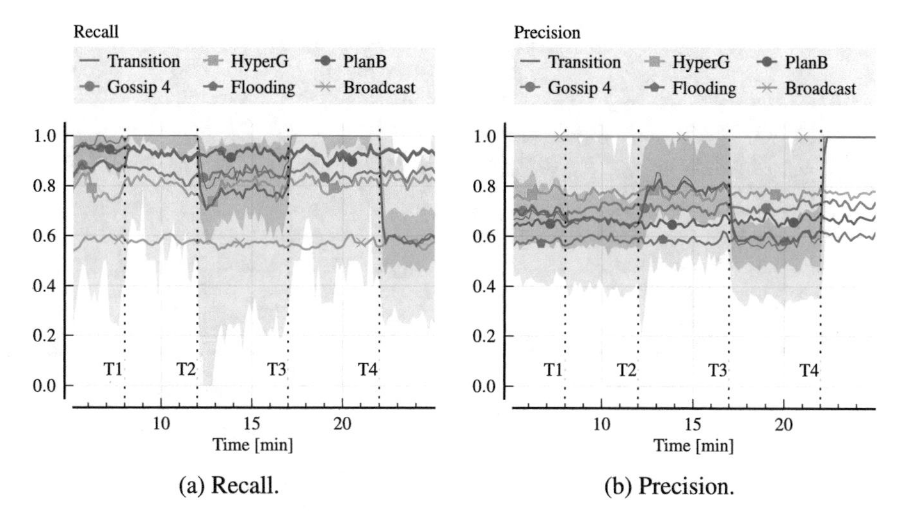

(a) Recall. (b) Precision.

Fig. 6.24 Recall and precision of transition-enabled dissemination mechanisms

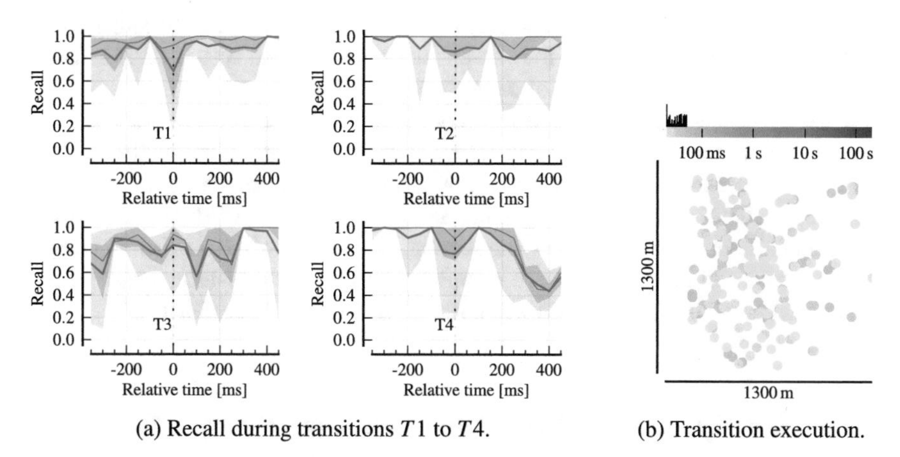

(a) Recall during transitions $T1$ to $T4$. (b) Transition execution.

Fig. 6.25 Recall before and after a transition execution

time frame of 50 ms. The plot indicates the location of a client when executing a transition and the duration measured from the point in time where the first instance of this transition is being executed in the network. This time frame is given by the configuration of the cellular network and its latency as introduced in Sect. 6.1.3.

Given that all clients receive the transition trigger within a rather confined time frame, the effect of message loss due to incompatible dissemination mechanisms is negligible. This is shown in Fig. 6.25a, where the recall is plotted during all four transitions. Although the effect of message loss can be observed for the transition $T1$, where the recall drops for approximately 50 ms, it is neither significant nor clearly visible for the remaining transitions. The drop for transition $T4$ at around 250 ms

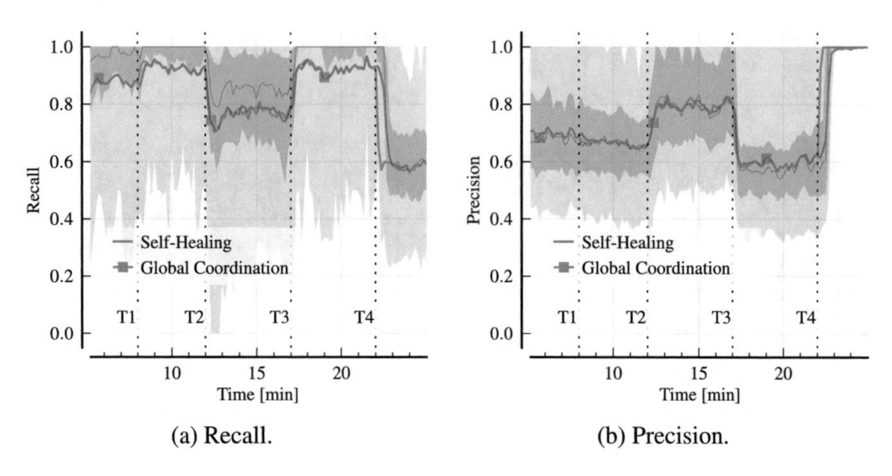

Fig. 6.26 Self-healing compared to centrally coordinated transitions

after the transition is executed is not caused by the transition itself but rather by the new dissemination mechanism: the one-hop dissemination via BROADCAST achieves an average recall of 0.6 only, as already shown in Fig. 6.24a.

As later discussed in Sect. 6.4, central coordination of transitions between dissemination mechanisms can cause significant control overhead. Additionally, in some scenarios, a central coordinator might not be available or network connectivity to all mobile clients cannot be guaranteed, as discussed in our work on pure ad hoc dissemination in [23]. In the following section, we study how the self-healing mechanism introduced in Sect. 4.3.3 can be utilized to spread a transition via an ad hoc network at no additional cost and without relying on central coordination.

6.3.4 Locally Initiated Transitions with Self-healing

The self-healing mechanism introduced in Sect. 4.3.3 can also be utilized to spread a transition decision within a MANET. In contrast to centralized coordination as discussed in the previous section, this process significantly reduces cellular traffic, as evaluated in Sect. 6.4.3 within the context of coexisting mechanisms. In this section, we characterize the spread of transitions in a MANET when relying solely on the self-healing mechanism. It is important to keep in mind that the self-healing mechanism is *passive* in that it does not spread the transition decision actively to neighbors. Instead, is relies on piggybacking to include information about the currently utilized mechanism within outgoing messages. Consequently, the transition decision is spread as a side-effect of the message forwarding procedure.

In this evaluation, transitions are again triggered at predefined times as previously discussed and summarized in Table 6.7. Instead of distributing the transition decision to all mobile clients via the cellular network, only a subset of clients is notified.

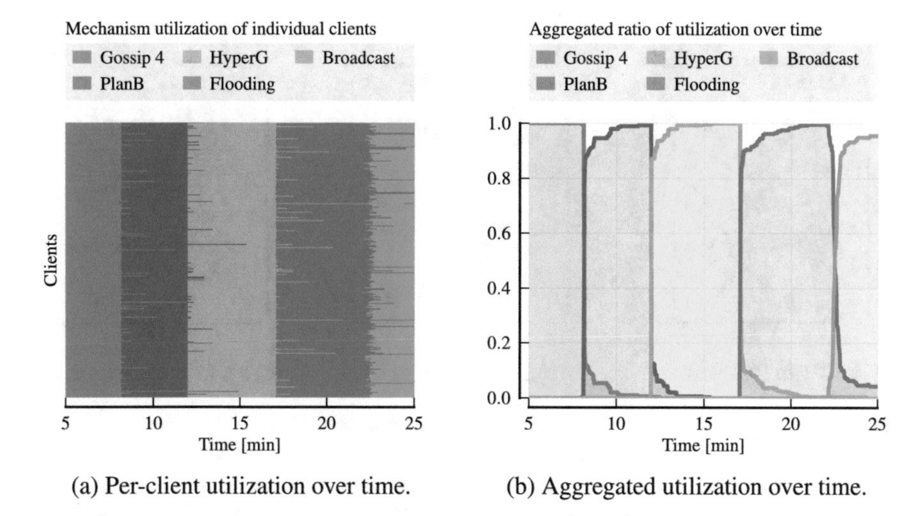

(a) Per-client utilization over time.

(b) Aggregated utilization over time.

Fig. 6.27 Mechanism utilization in case of transitions spreading via self-healing

The transition decision then spreads via the self-healing mechanism whenever a client forwards a message. Fig. 6.26 shows the recall and precision over time when transitions are triggered on a single randomly chosen client. For comparison, the behavior resulting from global coordination as discussed in the previous section is included in the plots. While the overall behavior is very much aligned to the case of global coordination, the effect of a transition is delayed. This effect is clearly visible after transitions $T1$ and $T4$ in Fig. 6.26a.

To study this effect in more detail, we plot the currently utilized mechanism over time for each client in Fig. 6.27. In Fig. 6.27a, each horizontal line corresponds to one single client. Clearly, most clients execute the transition at roughly the same time. However, a fraction of clients executes some of the transitions very late or not at all. For better readability, Fig. 6.27b shows the same data but grouped by mechanism. Here, it is clearly visible how a transition affects 90 % of the mobile clients on average, with the remaining clients taking significantly longer to execute the transition. This effect is especially severe for transition $T4$ to BROADCAST, as this dissemination mechanism does not forward messages more than one hop. Consequently, the transition decision induced by a single client at minute 22 has still not reached all clients at the end of the observation three minutes later.

To understand this behavior, we need to take the physical location of clients into account. In Fig. 6.28 we therefore plot the location of clients when executing a transition. The color indicates the delay between initiation and execution of a transition for each mobile client. A histogram of the respective delays is included for a better assessment of the characteristics of each individual transition. We can distinguish between three cases: (i) high initial delay and rapid spread Fig. 6.28a, (ii) low initial delay and rapid spread Fig. 6.28b, and (iii) slow spread Fig. 6.28c.

The reasons for the respective behavior are explained in the following.

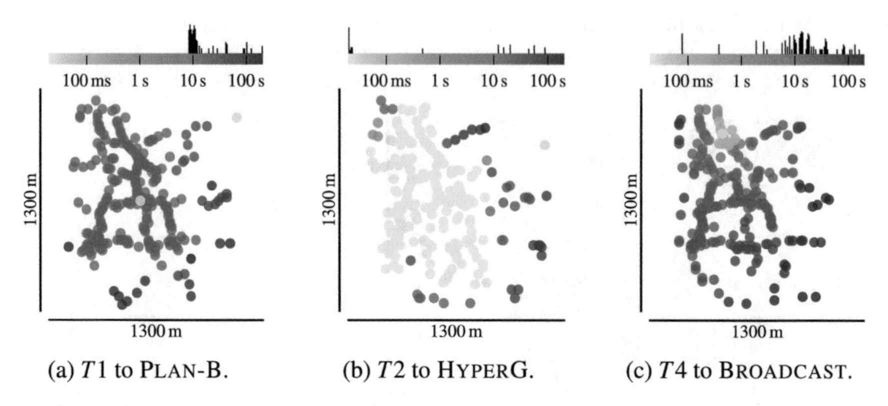

(a) $T1$ to PLAN-B. (b) $T2$ to HYPERG. (c) $T4$ to BROADCAST.

Fig. 6.28 Geographical spread of transitions with one initiator

For transition $T1$ to PLAN-B shown in Fig. 6.28a, we observe that only the initially triggered client in the center of the simulated area executes the transition immediately. However, after a delay of approximately ten seconds, the decision is rapidly spread to a large fraction of the remaining clients. Only a few clients in the outskirts of the simulated area do not receive the respective information, leading to increased delay of approximately one to two minutes. This is due to the fact that PLAN-B utilizes a distance-based hesitation mechanism, as discussed in Sect. 4.3.1 and [7]. Consequently, a client located within a rather densely populated area might not forward any messages at all. In this case, the transition decision can only spread if the client itself initiates a new message transmission, which can lead to a high initial delay, as observed in Fig. 6.28a. However, once the respective message is sent, it is rapidly spread to a large fraction of clients.

When executing transition $T2$ to HYPERG, one can observe a similar behavior, although this time the initial delay until further clients execute the transition is very low. This is caused by the active exchange of `Hello` messages in HYPERG [13] to determine the density of the current neighborhood. These messages are broadcast every 100 ms. As every outgoing mechanism-specific message, they contain piggy-backed information used by the self-healing mechanism as discussed in Sect. 4.3.3. Consequently, a transition is spread rapidly to nearby clients regardless of whether application payload needs to be forwarded or not.

For a single-hop dissemination mechanism like BROADCAST the achievable spread is clearly limited, as shown in Fig. 6.28c. As messages are not actively forwarded, a transition decision can only spread if a client actively publishes an event. Given that each broadcast reaches multiple clients, the probability of a client publishing an event after having executed the transition increases significantly after the initial broadcast took place. Consequently, as already discussed based on Fig. 6.27b, a large fraction of clients executes the transition within approximately 10–20 s.

A high initial delay is not necessarily bad with respect to message loss, given that only the initiator is unable to process incoming messages at this stage. However, once

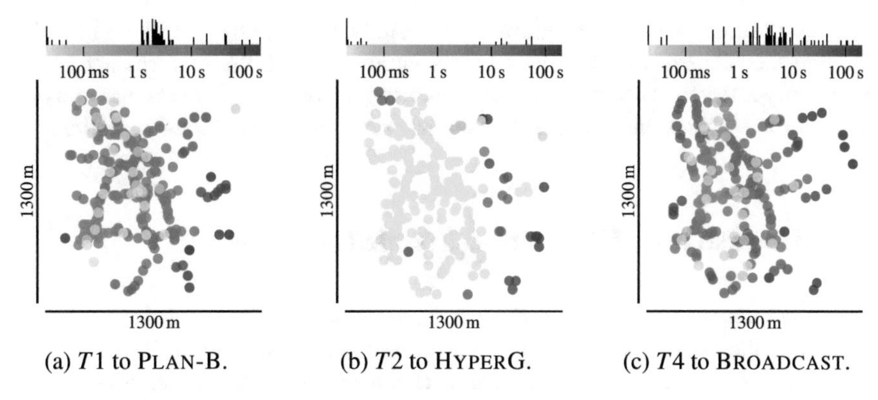

(a) $T1$ to PLAN-B. (b) $T2$ to HYPERG. (c) $T4$ to BROADCAST.

Fig. 6.29 Geographical spread of transitions with 30 initiators

the decision is spread to the remaining clients, it should reach a large fraction of the clients within a short time frame to prevent message loss. The previously discussed results for the recall shown in Fig. 6.26a support this statement. While the behavior is slightly delayed, the recall does not drop compared to a configuration with global coordination of transitions.

In hybrid configurations, and especially when considering the utilization of gateways, a transition decision could be injected at multiple clients. While this slightly increases the load on the cellular network, it comes at the advantage of a more rapid spread of the decision to the remaining nodes. This is visible in Fig. 6.29, where the transition decision is sent to 10 % of the clients initially, corresponding to 30 clients in our scenario. While the overall behavior of the individual mechanisms remains as previously discussed, the average delay for the spread of the decision decreases significantly compared to the results reported in Fig. 6.28. Results for FLOODING are omitted in this section for brevity, as the mechanism behaves similar to PLAN-B.

The results presented in this part of our evaluation motivate the utilization of hybrid modes for event dissemination. We show that pure ad hoc dissemination of events is not feasible within our scenario, given that the dissemination mechanisms are unable to cover the whole simulated area. Still, our results motivate the utilization of direct ad hoc dissemination for locally relevant events. The utilization of gateways further contributes to the goal of locality-aware event dissemination. By sending events only to a subset of clients within a region and then distributing them to the remaining clients in an ad hoc fashion, the cellular connection is offloaded significantly, while high recall is maintained.

Given that clients are not evenly distributed in our scenario, we want to utilize application-specific knowledge about attraction points, as already discussed for filter schemes. Our evaluation shows that transitions between dissemination mechanisms can be executed either in a centrally coordinated fashion or by relying on our self-healing mechanism. Regardless of the execution method, the performance is not negatively affected. This motivates the combined analysis of filter schemes and dissemination mechanisms and their transitions presented in the following section.

6.4 Coexistence of Transition-Enabled Mechanisms

In this section, we characterize system configurations that rely on coexisting mechanisms and the respective transitions, as discussed in Sect. 4.4 and further detailed in Sect. 5.3.2. Consequently, we combine mechanisms for location-based filtering and mechanisms for locality-aware event dissemination by executing the respective transitions based on application-specific knowledge about attraction points. With respect to the utilized filter schemes, the behavior remains the same as reported in Sect. 6.2.4 for the utilization of partial transitions for clients near attraction points. However, we vary the utilization of dissemination mechanisms for direct local dissemination and gateways according to the execution plans detailed in Sect. 5.3.2.

6.4.1 Direct Dissemination of Local Events Near Attraction Points

Transitions activating or deactivating a local dissemination mechanism are executed whenever a client enters or leaves the area covered by a channel in ATTRACT, as discussed in Sect. 5.3.2. This configuration of BYPASS.KOM is hereafter referred to as CX-AH. When activated, local dissemination is only used for *local events*, i. e., events that are published to a client's current location. Consequently, an effect should only be visible for the AR workload.

Figure 6.30 reports the performance characteristics of the CX-AH configuration. For comparison, we include a configuration where locally published events are *not* forwarded to the cloud. Instead, events are only distributed via the ad hoc protocol and the purely cloud-based event dissemination relying on RADIAL. In this configuration, we do not execute any transitions. The recall reported in Fig. 6.30a corresponds to the results reported for hybrid event dissemination in Sect. 6.3.2. However, as local dissemination is limited to the area defined by attraction points, the effects are less significant, as they only affect a subset of all mobile clients. As expected, there is no change in recall for the LBS workload, as clients do not issue any local events.

The precision reported in Fig. 6.30b drops due to the utilization of the channel-based filter scheme ATTRACT for both workloads, as previously discussed in Sect. 6.2.4. Given that CX-AH behaves as the previously discussed hybrid configurations for clients within reach of an attraction point, we do not expect any offloading effect. This is supported by the results reported in Fig. 6.30c: the overall cellular traf-

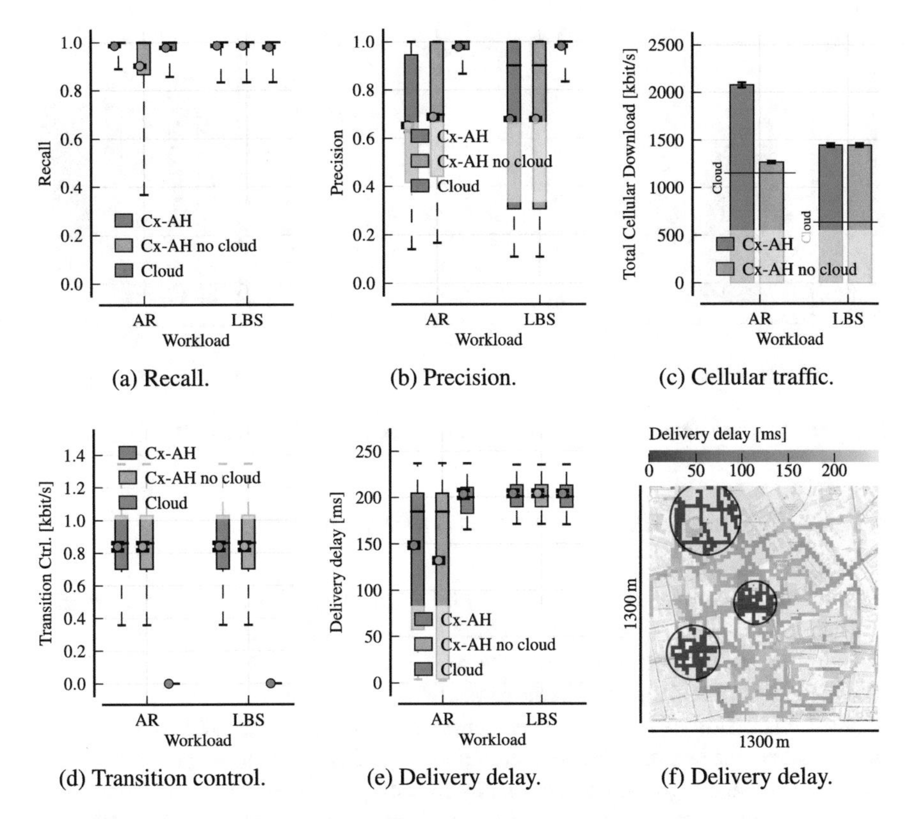

(a) Recall. (b) Precision. (c) Cellular traffic.

(d) Transition control. (e) Delivery delay. (f) Delivery delay.

Fig. 6.30 Performance of ad hoc dissemination around attraction points

fic increases due to the utilization of the channel-based filter scheme and its decreased precision.

To assess whether the increase in traffic compared to the purely cloud-based approach is caused by the filter scheme or by the coordination of transitions, we report the traffic caused by transition coordination in Fig. 6.30d. Compared to the overall traffic reported in Fig. 6.30c, the coordination of transitions accounts for less than 0.1 % of the cellular traffic in both configurations of CX-AH. Still, the average delivery delay is decreased from approximately 200 ms to about 150 ms if ad hoc dissemination mechanisms are utilized nearby attraction points, as reported in Fig. 6.30e.

The spatial distribution of the delivery delay is more interesting than average values, given that local dissemination mechanisms are only used nearby attraction points. Therefore, we report the delivery delay depending on the location of a client in the AR workload in Fig. 6.30f. The figure shows that local ad hoc dissemination is limited to the channels within ATTRACT, with an average delivery delay of below 50 ms. For locations that are not covered by a channel the delivery delay remains at around 200 ms, as also reported for the cloud-only configuration.

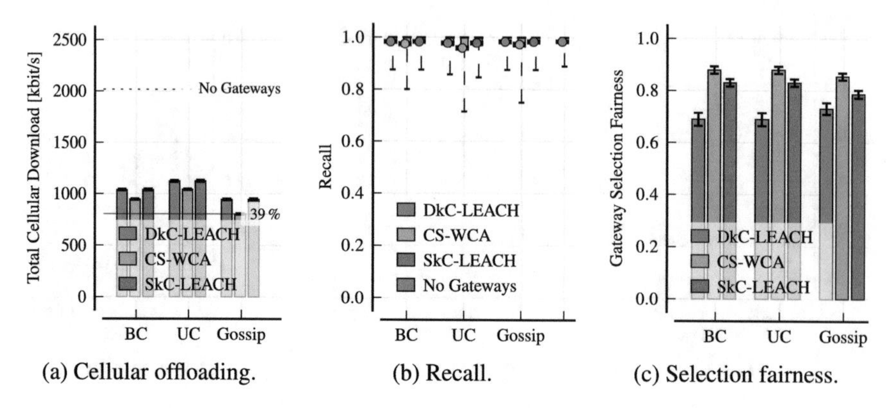

(a) Cellular offloading. (b) Recall. (c) Selection fairness.

Fig. 6.31 Achieved offloading, recall, and fairness characteristics (AR workload)

6.4.2 Utilizing Gateways Nearby Attraction Points

To offload the cellular connection, we utilize gateways in the area covered by a channel in ATTRACT, as discussed in Sect. 5.3.2. We utilize three distinct gateway selection algorithms (DKC-LEACH, CS-WCA, and SKC-LEACH) as representative examples of the algorithms included in the respective module published in [25]. For a more detailed discussion of the gateway selection algorithms, please refer to Sect. 4.3.4. As motivated, we evaluate different combinations of dissemination mechanisms and gateway selection algorithms, specifically focusing on the impact of multi-hop versus single-hop dissemination. We utilize GOSSIP with a forwarding probability of $p = 0.4$ as multi-hop dissemination mechanism, and UNICAST and BROADCAST as single-hop mechanisms. UNICAST can be utilized as the gateway selection procedure returns an assignment of clients to gateways, as discussed in Sect. 4.3.4. However, in this case, the contact information for the respective clients need to be forwarded to the gateway, leading to increased message sizes and, thus, additional cellular communication overhead. This effect is visible in Fig. 6.31a when comparing the cellular traffic of the configurations using BROADCAST and UNICAST regardless of the utilized gateway selection algorithm. Figure 6.31a additionally shows that by utilizing gateways within ATTRACT channels, the cellular traffic is reduced by up to 64 % compared to a configuration without gateways.

The chosen dissemination mechanism does not have a significant impact on the recall for DKC-LEACH and SKC-LEACH, shown in Fig. 6.31b. The recall of CS-WCA is slightly worse compared to the other algorithms and a configuration without gateways. Before discussing the reason for this decrease, we briefly discuss the fairness of the individual gateway selection algorithms, shown in Fig. 6.31c. The gateway selection fairness for all N clients is calculated using the fairness index defined by Jain in et al. [8]:

$$J(x) = \frac{\left(\sum_{i=1}^{N} x_i\right)^2}{N \sum_{i=1}^{N} x_i^2}.$$

Here, x_i is a counter of how often client i is chosen as a gateway. An algorithm with $J(x)$ closer to 1 distributes load across mobile clients more evenly.

The selection fairness of individual algorithms does not depend on the utilized dissemination scheme. The difference between the single-hop dissemination mechanisms and GOSSIP is caused by the configuration of the gateway selection algorithms: if a multi-hop dissemination mechanism is used, less gateways are requested by BYPASS.KOM. This is controlled through the parameter k as introduced in Sect. 5.3.2. In BYPASS.KOM, k is calculated based on the number of clients $|U|$ subscribed to the respective channel as $k = \lceil |U| \times k' \rceil$. For single-hop dissemination mechanisms we set $k' = 0.4$, corresponding to 40 % of all subscribers being used as gateways. In case of multi-hop dissemination, k' is reduced to 0.2. Consequently, the selection fairness behaves slightly different, as less gateways are selected in this case. In general, given that we only consider how often a client acts as a gateway, the stochastic approaches CS-WCA and SKC-LEACH achieve slightly higher selection fairness.

In real-world applications, the selection fairness is usually not a suitable indicator to compare selection algorithms. Instead, one wants to consider the available battery power or the quality of the cellular network connection when determining suitable gateways. Here, more elaborate selection algorithms come into play, as discussed for offloading in transition-enabled monitoring systems in [25]. Figure 6.32 aids in understanding the previous results by showing spatial characteristics of the gateway selection algorithms. Figure 6.32a shows the average number of subscribers interested in an event as spatial metric, regardless of the utilized filter scheme. This corresponds to the ground truth resulting from the circular AoI defined by the application when issuing a location-based subscription.

In comparison, the number of utilized gateways for events issued at the respective locations is shown for SKC-LEACH in Fig. 6.32b and CS-WCA in Fig. 6.32c in case of single-hop dissemination via BROADCAST. The selection algorithm CS-WCA selects on average 20 % less gateways compared to SKC-LEACH, especially for the large area of the top left attraction point. This observation is also backed by the aggregated number of utilized gateways per event in Fig. 6.32d. Additionally, as a consequence of the transition execution plan discussed in Sect. 5.3.2, gateways are only utilized for events that are distributed to clients within a channel in ATTRACT. The area is larger than the radius of the individual attraction points, as we add the average subscription radius as previously discussed in Sect. 6.2.4.

Figure 6.32e, f report the number of delivery attempts for events based on their target location for SKC-LEACH and CS-WCA, respectively. A larger value indicates redundant coverage of a client by multiple gateways, as we utilize the single-hop BROADCAST dissemination mechanism. Due to the higher number of gateways being select by SKC-LEACH, events are delivered up to four times to clients within proximity of an attraction point, as shown in Fig. 6.32e.

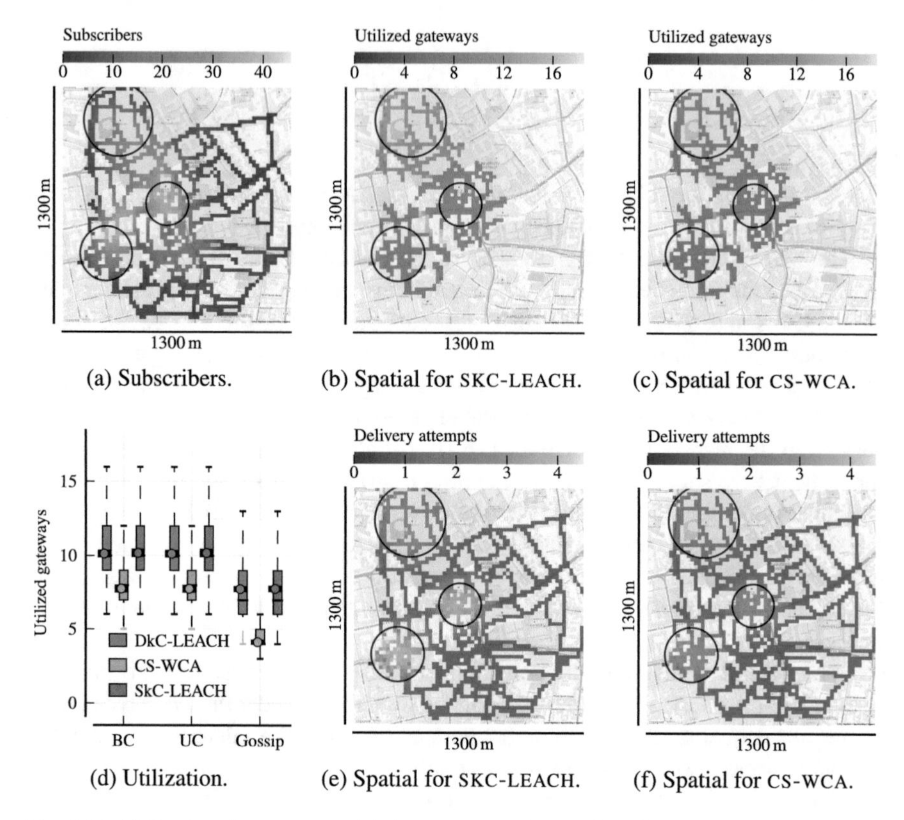

(a) Subscribers. (b) Spatial for SKC-LEACH. (c) Spatial for CS-WCA.

(d) Utilization. (e) Spatial for SKC-LEACH. (f) Spatial for CS-WCA.

Fig. 6.32 Utilization of gateways and delivery attempts (AR workload)

This is due to the fact that we determine the preferred number of gateways (the parameter k) solely based on the number of subscribers and not on their average density. The clustering-based algorithm CS-WCA takes client density into account (as it has a direct impact on the size of a cluster), thereby reducing the number of gateways and, consequently, the delivery attempts compared to SKC-LEACH. A simple solution to include an estimate of the client density into the calculation of k is to utilize the radius r of an attraction point or to chose a gateway selection algorithm that already takes the node density into account. We propose an alternative approach in Sect. 5.3.2, relying on transitions to adapt the dissemination mechanism used by gateways based on the client density. The characteristics of this approach are discussed in the following section.

6.4.3 *Executing Mechanism Transitions Based on Client Density*

As proposed in Sect. 5.3.2, execution plans for transitions can be determined based on the observed client density to further adapt the system to application-specific dynamics. In this section, we discuss the resulting characteristics of the system, focusing on the utilization of the respective mechanisms under varying client density. Consequently, this evaluation serves as a proof of concept for the applicability of our transition-specific contributions to the SIMONSTRATOR.KOM platform. It is not intended as a performance evaluation of the resulting system, given the rather arbitrary choice of our execution rules presented in Sect. 5.3.2.

We adapt the dissemination mechanisms used for the direct dissemination of local events and for gateways as discussed in Sect. 5.3.2. The respective transitions are executed if the observed client density ρ at an attraction point crosses the threshold δ. In this evaluation, we calculate ρ using the radius of an attraction point r in meters and the number of clients subscribed to the respective channel, $|U|$, as follows:

$$\rho = |U| \times \frac{10{,}000\,\mathrm{m}^2}{r^2}\;.$$

Thereby, ρ normalizes the number of subscribers to the area of an attraction point with $r = 100\,\mathrm{m}$. We report the system behavior for different values of δ.

In contrast to previous evaluations, we vary the total number of clients during the experiment to trigger the density changes. To this end, we utilize a churn model that starts with 100 clients. After 8 min of simulated time, the model linearly increases the number of active clients over a time span of 7 min, until finally reaching the total number of 297 active clients as in our default setup. The number of clients remains constant for another 3 min, before the model returns to 100 clients within a time span of 7 min. Regardless of whether a client is active or not, it moves according to the mobility models previously discussed.

Figure 6.33 shows the performance characteristics of our system with varying δ over time. For reference, the number of currently active clients within the previously discussed churn model is included in each plot. The results for recall and precision show that the system behaves stable regardless of the configuration of δ or the execution of our proposed transitions. As expected, the number of gateways used to distribute an event increases with the density of clients, as more clients are active in the area around attraction points.

To validate the correct operation of our execution plans, we plot the location of transitions executions affecting the gateway dissemination mechanism in Fig. 6.34a, b. In both cases, transitions are only executed on clients that are currently in the area assigned to an attraction point. For comparison, the locations of filter scheme transitions between ATTRACT and RADIAL are shown in Fig. 6.34c, d. Filter scheme transitions are only executed when clients enter or leave the area assigned to an attraction point. There is one exception: clients that join the system due to our churn

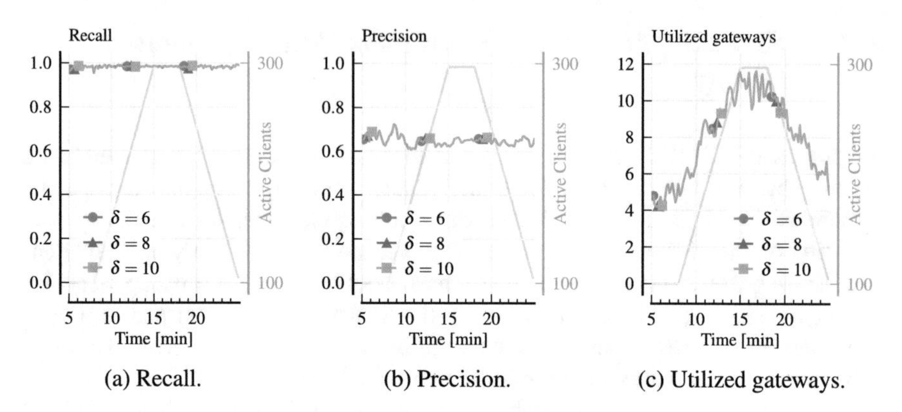

(a) Recall. (b) Precision. (c) Utilized gateways.

Fig. 6.33 Performance over time for varying δ; AR workload and DA scenario

model while being near an attraction point also execute the respective transition. This is visible in comparison to the results reported in Sect. 6.2.4 for a constant number of clients. Still, compared to transitions between gateway dissemination mechanisms, the execution of transitions between filter schemes mostly occurs at the border of the area assigned to an attraction point.

In the Mannheim scenario as shown in Fig. 6.34b, clients near the attraction point on the top left of the map *exclusively* utilize UNICAST as gateway dissemination mechanism. The client density at this attraction point, ρ, never exceeded $\delta = 8$. Near all other attraction points, both mechanisms are utilized by clients. Figure 6.35 reports the corresponding mechanism utilization over time for different values of δ in the DA scenario. A ratio of 1.0 would indicate that all 297 clients are currently utilizing the respective mechanism. In all three cases, approximately 35 % of all clients are subscribed to an attraction point channel. Depending on the value of δ, at some point in time all clients use BROADCAST as their gateway dissemination mechanism. Once the client density starts decreasing again (after minute 18), this effect is reversed and clients are again instructed to use UNICAST around most attraction points. The discrete steps visible in the plots correspond to an adaptation of the utilized mechanism for all clients in an attraction point. For $\delta = 10$ one can observe oscillations between both mechanisms. In this case, the client density at an attraction point oscillates around the value of δ, leading to the observed behavior.

Notably, these oscillations do not have any visible effect on the performance, as reported in Fig. 6.33. This is due to the fact that oscillations occur at relatively low frequencies compared to the time it takes to distribute an execution plan and execute the respective transition on all affected clients. For reference, a global transition affecting all clients is executed within approximately 50 ms counted from the execution at the first client, as reported in Sect. 6.3.3. Consequently, the effect of locally mismatching dissemination mechanisms is not of concern in this scenario. Nevertheless, to avoid oscillations and the associated control overhead, one could add additional hysteresis to the respective execution plan introduced in Sect. 5.3.2.

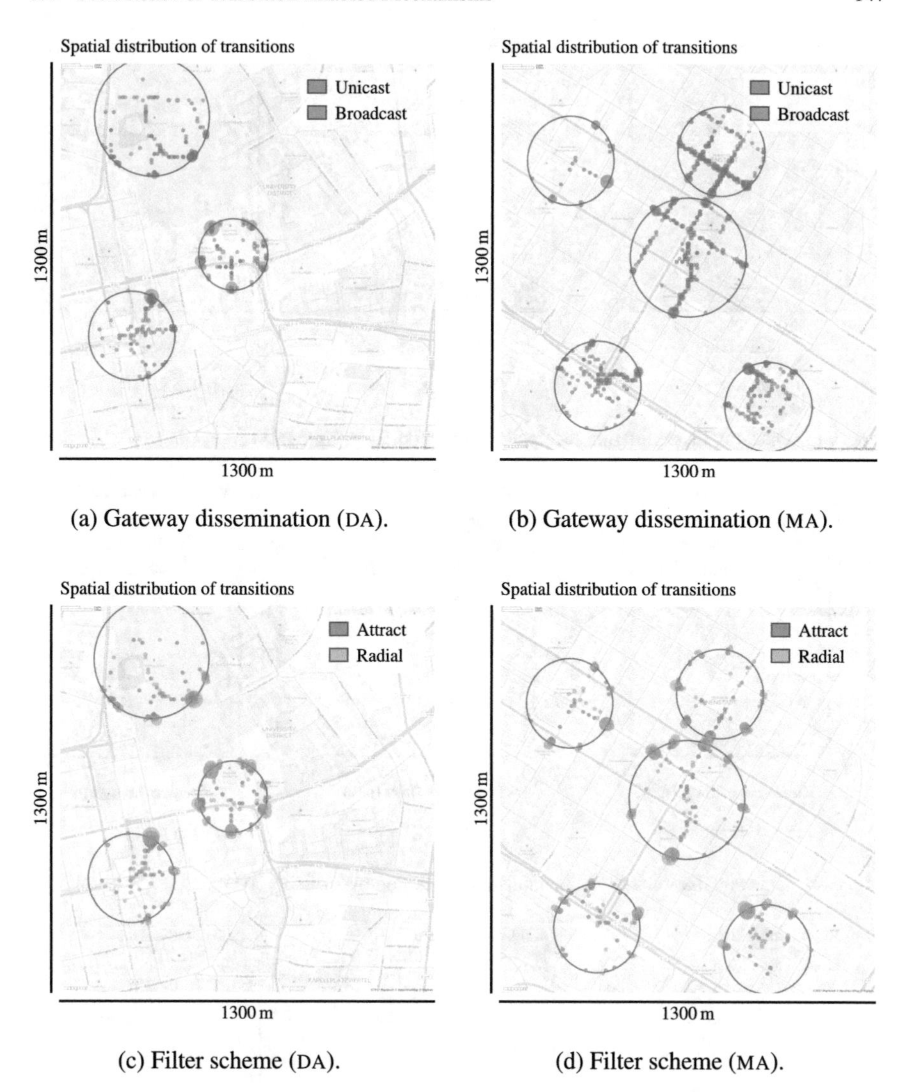

(a) Gateway dissemination (DA).

(b) Gateway dissemination (MA).

(c) Filter scheme (DA).

(d) Filter scheme (MA).

Fig. 6.34 Client locations when executing transitions with $\delta = 8$

Individual mechanism utilization per host is reported in Fig. 6.36. With increasing δ, the utilization shifts from BROADCAST to UNICAST. The duration neither BROADCAST nor UNICAST is utilized remains constant regardless of δ, as it is solely determined by the mobility model. As the reported results utilize the same random seed for the mobility model, this behavior does not differ.

We report the results for the MA scenario in Fig. 6.37 for comparison. Due to the larger number of attraction points, the overall behavior over time is more diverse than in the DA scenario. Still, discrete jumps in mechanism utilization over time are

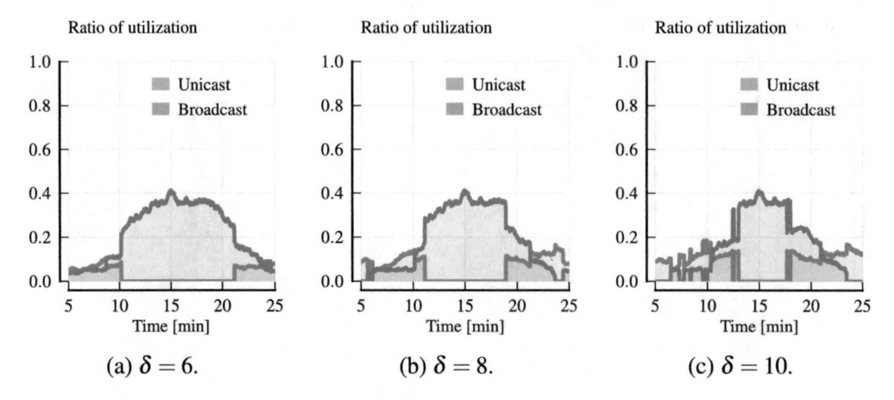

Fig. 6.35 Gateway dissemination mechanism utilization over time (DA)

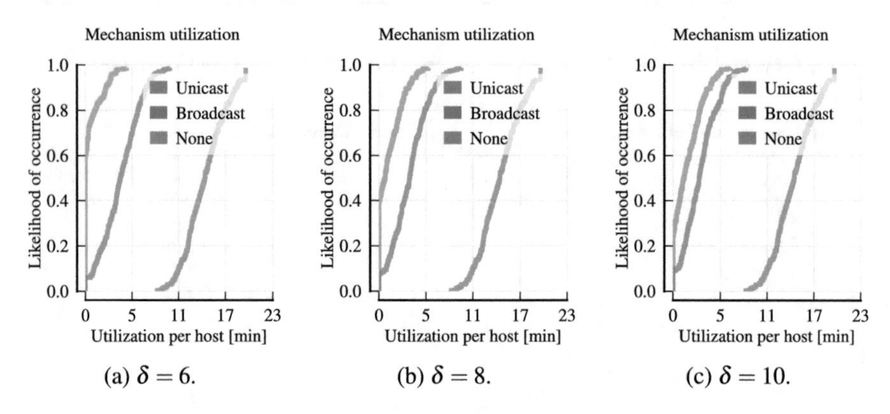

Fig. 6.36 Gateway dissemination mechanism distribution over clients (DA)

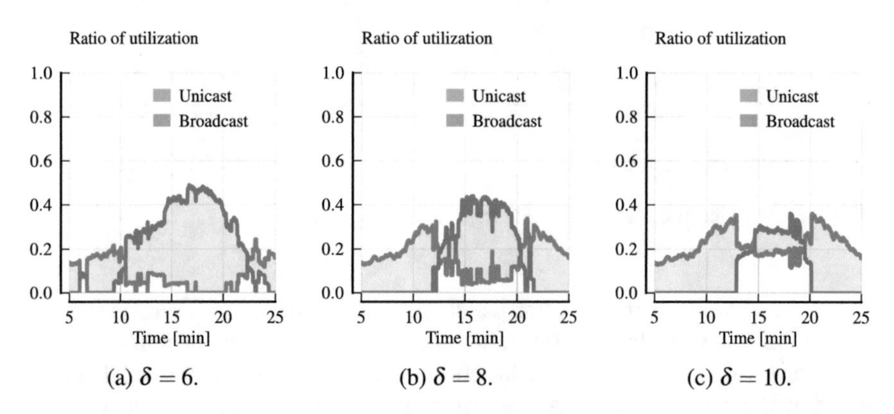

Fig. 6.37 Gateway dissemination mechanism utilization over time (MA)

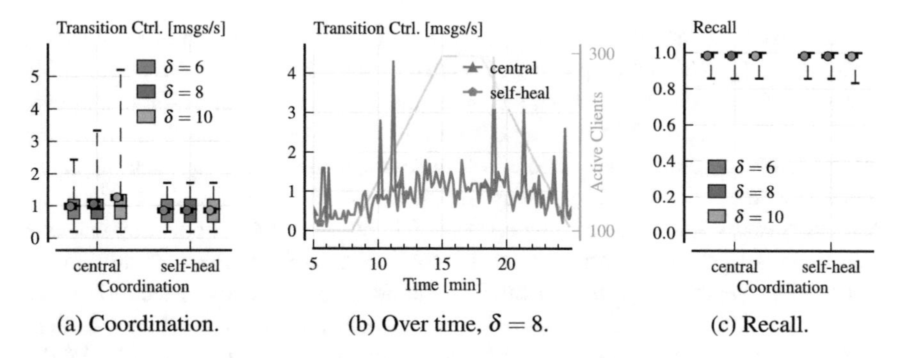

(a) Coordination. (b) Over time, $\delta = 8$. (c) Recall.

Fig. 6.38 Central coordination of transitions compared to self-healing (DA)

visible in all cases. As shown in the spatial distribution of transition reported for $\delta = 8$ in Fig. 6.34b, at some attraction points no transition between dissemination mechanisms is triggered at all. For larger values of δ as reported in Fig. 6.37c, this is the case for an even larger fraction of the attraction points.

In the following, we discuss the coordination overhead resulting from the dissemination of transition execution plans to mobile clients. We further show the potential of using the self-healing mechanism as proposed in Sect. 5.3.2 to reduce the overhead of central coordination. Figure 6.38a reports the number of messages sent to control transitions for different values of δ, comparing central coordination with self-healing. We are not interested in the absolute message sizes, as it significantly depends on how the transition execution plan is encoded and compressed. This technical aspect is not addressed in our thesis. In the self-heal approach, the execution plans used to adapt the local dissemination mechanisms is not actively disseminated. Clients entering the area of an attraction point are instructed to use the new mechanism. The decision then spreads as a side effect of event dissemination due to our self-healing mechanism proposed in Sect. 4.3.3.

As expected, the self-heal approach leads to transition coordination traffic that is independent of the chosen value for δ. In contrast, if the decision is to be distributed to all clients that are currently subscribed to an attraction point channel, the increasing number of transitions for higher δ leads to increased coordination overhead whenever the local dissemination mechanism is to be changed.

This is visible in Fig. 6.38b, where the coordination traffic is reported over time for $\delta = 8$. Both coordination schemes behave identical in terms of triggering transitions for clients that enter or leave the area around an attraction point. However, for central coordination, significant spikes are observed whenever all clients within a channel have to adapt their local dissemination mechanism. Using the self-healing mechanism does not lead to a degradation of the overall performance, as reported in Fig. 6.38c. Only for $\delta = 10$, a slight degradation is visible for the reported 2.5th percentile. This effect can be mitigated if oscillations are avoided.

In this part of our evaluation, we discussed the utilization of mechanism transitions to adapt to application-specific workloads and mobility characteristics. To this end, we evaluated the utilization of direct ad hoc dissemination of local events within vicinity of attraction points in Sect. 6.4.1. We showed that the dissemination delay is reduced significantly. This enables more interactive applications and can, for example, be utilized to exchange information that reduces the delay of visual effects caused by other users' actions within an augmented reality application. Combined with the utilization of gateways as evaluated in Sect. 6.4.2, the load on the cellular network can be reduced by up to 64 % in our scenario, without any negative effects on the overall performance of the publish/subscribe system. Finally, in the last section, we evaluated the potential of adapting the respective execution plans and, thereby, the choice of dissemination mechanisms, based on the observed client density around attraction points. We further discussed the overhead caused by the coordination of transitions and demonstrated how the self-healing mechanism can be utilized to significantly reduce the cost of this process.

Overall, we successfully demonstrated the combined and adaptive utilization of mechanisms for location-based filtering and locality-aware dissemination of events through the execution of mechanism transitions.

References

1. Abhayawardhana VS, Wassell IJ, Crosby D, Sellars MP, Brown MG (2005) Comparison of empirical propagation path loss models for fixed wireless access systems. In: Proceedings IEEE vehicular technology conference (VTC). vol 1. IEEE, pp 73–77
2. Bai F, Sadagopan N, Helmy A (2003) IMPORTANT: a framework to systematically analyze the impact of mobility on performance of routing protocols for Adhoc networks. In: Proceedings IEEE international conference on computer communications (INFOCOM). vol 2. IEEE, pp 825–835
3. Baumann P, Klaus J, Richerzhagen B, Kleiminger W, Santini S (2015) The UbiDCC data set: collecting Wi-Fi and bluetooth scans during a large-scale conference. In: Proceedings 1st workshop on context sensing and activity recognition (CSAR). ACM, pp 25–30
4. Bianchi G (2000) Performance analysis of the IEEE 802.11 distributed coordination function. IEEE J Sel Areas Commun 18(3):535–547
5. Dorine C, Winnie Daamen D, Hoogendoorn SP (2013) State-ofthe- art crowd motion simulation models. Transp Res Part C: Emerg Technol 37:193–209
6. Hasan S, Zhan X, Ukkusuri SV (2013) Understanding urban human activity and mobility patterns using large-scale location-based data from online social media. In: Proceedings ACM SIGKDD international workshop on urban computing. ACM, p 6
7. Holzer A, Vessaz F, Pierre S, Garbinato B (2011) PLAN-B: Proximity-based lightweight adaptive network broadcasting. In: Proceedings international symposium on network computing and applications (NCA). IEEE, pp 265–270
8. Jain R, Chiu D-M, Hawe WR (1984) A quantitative measure of fairness and discrimination for resource allocation in shared computer system. vol 38. Eastern Research Laboratory, Digital Equipment Corporation Hudson, MA
9. Jayarajah K, Lee Y, Misra A, Balan RK (2015) Need accurate user behaviour?: pay attention to groups! In: Proceedings ACM international joint conference on pervasive and ubiquitous computing. ACM, pp 855–866
10. Johnson DB, Maltz DA (1996) Dynamic source routing in ad hoc wireless networks. In: Mobile computing. Springer, pp 153–181

11. Kaup F, Fischer F, Hausheer D (2017) Measuring and predicting cellular network quality on trains. In: Proceedings international conference on networked systems (NetSys). IEEE, pp. 1–8
12. Kaup F, Jomrich F, Hausheer D (2015) Demonstration of network coverage—a mobile network performance measurement app. In: Proceedings international conference on networked systems (NetSys). IEEE
13. Khelil A, Josê Marrôn P, Becker C, Rothermel K (2007) Hypergossiping: a generalized broadcast strategy for mobile ad hoc networks. Ad Hoc Networks 5(5):531–546
14. Knoblauch R, Pietrucha M, Nitzburg M (1996) Field studies of pedestrian walking speed and start-up time. Transp Res Record: J Transp Res Board 1538:27–38
15. Kurkowski S, Camp T, Colagrosso M (2005) MANET simulation studies: the incredibles. ACM SIGMOBILE Mobile Comput Commun Rev 9(4):50–61
16. Kurkowski S, Navidi W, Camp T (2007) Constructing manet simulation scenarios that meet standards. In: Proceedings IEEE international conference on mobile adhoc and sensor systems (MASS). IEEE, pp 1–9
17. MacCartney GR, Zhang J, Nie S, Rappaport TS (2013) Path loss models for 5G millimeter wave propagation channels in urban microcells. In: Proceedings IEEE global communications conference (GLOBECOM). IEEE, pp. 3948–3953
18. Michelinakis F, Bui N, Fioravantti G, Widmer J, Kaup F, Hausheer D (2015) Lightweight mobile bandwidth availability measurement. In: Proceedings IFIP networking conference (IFIP Networking). IEEE, pp 1–9
19. Michelinakis F, Bui N, Fioravantti G, Widmer J, Kaup F, Hausheer D (2016) Lightweight capacity measurements for mobile networks. Comput Commun 84:73–83
20. Richerzhagen B, Stingl D, Hans R, Groß C, Steinmetz R (2014) Bypassing the cloud: peer-assisted event dissemination for augmented reality games. In: Proceedings 14th IEEE conference on peer-to- peer computing (P2P), pp 1–10
21. Richerzhagen B, Stingl D, Rückert J, Steinmetz R (2015) Simonstrator: simulation and prototyping platform for distributed mobile applications. In: Proceedings 8th international conference on simulation tools and techniques (SIMUTOOLS). ACM, pp 99–108
22. Richerzhagen B, Schiller M, Lehn M, Lapiner D, Steinmetz R (2015) Transition-enabled Event dissemination for pervasive mobile multiplayer games. In: Proceedings 16th international symposium on a world of wireless, mobile and multimedia networks (WoWMoM). IEEE
23. Richerzhagen B, Wagener A, Richerzhagen N, Hark R, Steinmetz R (2016) A framework for publish/subscribe protocol transitions in mobile crowds. In: Proceedings 10th international conference on autonomous infrastructure, management and security (AIMS). IFIP, pp 1–14
24. Richerzhagen B, Richerzhagen N, Zobel J, Schönherr S, Koldehofe B, Steinmetz R (2016) Seamless transitions between filter schemes for location-based mobile applications. In: Proceedings 41st IEEE conference on Local computer networks (LCN), pp 1–9
25. Richerzhagen N, Richerzhagen B, Walter M, Stingl D, Steinmetz R (2016) Buddies, not enemies: fairness and performance in cellular offloading. In: Proceedings 17th international symposium on a world of wireless, mobile and multimedia networks (WoWMoM). IEEE, pp 1–9
26. Richerzhagen N, Richerzhagen B, Stingl D, Steinmetz R (2017) The human factor: a simulation environment for networked mobile social applications. In: Proceedings international conference on networked systems (NetSys). IEEE, pp 1–8
27. Riley GF, Henderson TR (2010) The ns-3 network simulator. In: Modeling and tools for network simulation, pp 15–34
28. Stingl D (2014) Decentralized monitoring in mobile ad hoc networks—provisioning of accurate and location-aware monitoring information. PhD thesis. TU Darmstadt
29. Stingl D, Richerzhagen B, Zöllner F, Gross C, Steinmetz R (2013) PeerfactSim.KOM: take it back to the streets. In: Proceedings international conference on high performance computing and simulation (HPCS). IEEE, pp 80–86
30. Van WG, Toll FC, Geraerts R (2012) Real-time densitybased crowd simulation. Comput Anim Virtual Worlds 23(1):59–69
31. Vu L, Nguyen P, Nahrstedt K, Richerzhagen B (2015) Characterizing and modeling people movement from mobile phone sensing traces. Pervasive Mobile Comput 17:220–235

Chapter 7
Summary, Conclusions, and Outlook

To conclude our work, we summarize the content of the previous chapters and state our main contributions in the following. We then draw conclusions based on our obtained results. Finally, we discuss open issues and potential future work.

7.1 Summary of the Thesis

In Chap. 1, we described the challenges for a communication system that result from location-based mobile social applications. We motivated the utilization of publish/subscribe as communication paradigm to address the interaction among users typical for these applications, described in detail in Chap. 2. To address dynamics arising from user mobility and application-specific attraction points, we motivated the design concept of mechanism transitions. We studied existing mechanisms for location-based filtering and locality-aware dissemination and proposed their combined utilization in Chap. 3. Additionally, we discussed related approaches to adaptivity in publish/subscribe and studied the utilization of mechanism transitions in other application domains. Based on our analysis of the state of the art, we presented the following contributions in our thesis.

7.1.1 Contributions

BYPASS.KOM, introduced in Chap. 4, is a framework to study the potential of mechanism transitions for location-based filtering and locality-aware dissemination of events, constituting our *first contribution*. In this framework, we presented a methodology and, consequently, a design for transitions between filter schemes for location-based publish/subscribe. We proposed an encapsulation method for filter schemes,

B. Richerzhagen, *Mechanism Transitions in Publish/Subscribe Systems*,
Springer Theses, https://doi.org/10.1007/978-3-319-92570-7_7

consisting of client and broker components, a scheme-specific context update protocol, and the respective storage for context information at the broker. Thereupon, we discussed the integration of context-based and channel-based filter schemes and designed transitions to switch between filter schemes of both types at runtime. Our design includes a state transfer mechanism to address the challenge of seamless execution of transitions. In contrast to existing work on location-based publish/subscribe, our design is not limited to parameter adaptations within a filter scheme. Instead, existing and upcoming filter schemes can be utilized by integrating them as transition-enabled mechanisms into BYPASS.KOM.

In addition to location-based filtering, we addressed locality-aware event dissemination by enabling transitions between local dissemination mechanisms in BYPASS.KOM. We extended and applied our methodology for the encapsulation of filter schemes to ad hoc dissemination mechanisms. Thereupon, we proposed hybrid modes of operation that specifically address distinct communication interfaces available on today's mobile devices. We explicitly distinguished between local events (being relevant within vicinity of their publisher) and generic location-based events targeted at arbitrary locations. This enabled us to benefit from geofencing and locally confined dissemination mechanisms. We further proposed a self-healing mechanism to resolve situations where mobile clients utilize different dissemination mechanisms as a potential consequence of transitions. Existing approaches to locality-aware event dissemination focus either solely on local ad hoc networks, or are limited to the organization of distributed broker networks. Additionally, they do not consider the peculiarities of location-based publish/subscribe, leading to unnecessary overhead in the dissemination of events. In contrast, our approach of transition-enabled event dissemination mechanisms combined with hybrid modes of operation allowed us to tackle the challenge of efficient utilization of heterogeneous networks.

Finally, we proposed the combined utilization of mechanisms for location-based filtering and locality-aware event dissemination in BYPASS.KOM depending on application-specific attraction points to address the mobility characteristics of mobile social applications. To coordinate transitions between coexisting mechanisms, we modeled the respective transitions through execution plans. We introduced self-transitions as a generic concept to adapt a mechanism's state or configuration parameters. Thereby, we can utilize two distinct concepts for adaptivity, (parameter) reconfiguration, and mechanism transitions in a mechanism-independent fashion. Current adaptive publish/subscribe middleware either does not offer the ability to completely exchange mechanisms at all, or it offers only global reconfiguration.

With our *second contribution*, we generalized our findings by contributing transition-specific design abstractions and components to the SIMONSTRATOR.KOM platform, as discussed in Chap. 5. Further, we contributed mobility and network models for the scenario of mobile social applications. The SIMONSTRATOR.KOM platform and our transition-specific contributions constitute the foundation for the evaluation and characterization of mechanism transitions presented in Chap. 6.

7.1.2 Conclusions

We conducted an extensive evaluation to assess the effects of transitions between different mechanisms on a publish/subscribe system. We showed that our transition-enabled system enables the utilization of individual mechanisms with their respective performance and cost characteristics. It correctly executes transitions between mechanisms, both for location-based filtering and locality-aware event dissemination. Consequently, by integrating the respective mechanisms into our framework, the overall system can adapt to a broad range of conditions, as envisioned in the Collaborative Research Centre "MAKI". By transferring state during a transition, communication between clients and the broker during initialization of the target mechanism is avoided. We compared the behavior of location-based filter schemes with and without state transfer, showing that the mechanism substantially contributes to a seamless execution of transitions. We further evaluated our concept of partial transitions that switch the filter scheme for a subset of clients, motivated by the need for local adaptation to application-specific workload and mobility characteristics. We showed that a broker can execute a partial transition without requiring any client-side modifications. We argue that partial transitions constitute an important foundation for systems that need to adapt to heterogeneous characteristics.

We showed that pure ad hoc dissemination of events is not feasible for location-based mobile social applications, given that such dissemination mechanisms are unable to cover larger distances. Still, our results motivate the utilization of direct ad hoc dissemination for locally relevant events. We evaluated the combination of location-based filter schemes and locality-aware event dissemination mechanisms, proposing the utilization of gateway selection algorithms to further offload the cellular connection. Our evaluation results show that locally confined partial transitions around application-specific attraction points enable the publish/subscribe system to adapt to heterogeneous requirements, while at the same time maintaining high performance. We show how the self-healing mechanism proposed for locality-aware dissemination schemes can further reduce the coordination cost of transitions.

Lastly, we showed the applicability of our proposed generic design concepts for transition-enabled communication systems. By applying the abstractions contributed to the SIMONSTRATOR.KOM platform, transitions between publish/subscribe mechanisms are coordinated and executed in a mechanism-independent fashion. Most notably, the proposed concept of self-transitions enables fine-grained reconfiguration of the utilized transition-enabled mechanisms in a unified fashion. We demonstrated the coordinated execution of total transitions, partial transitions, and self-transitions through execution plans. We evaluated the adaptation of the execution plans to environmental parameters such as the currently observed client density, further demonstrating the potential of mechanism transitions as a generic way to realize adaptivity in communication systems.

7.2 Outlook

The results presented in this thesis motivate the evaluation of transition-enabled publish/subscribe systems in other application scenarios, where high adaptivity to the respective application-specific characteristics is required. Especially in the automotive sector, location-based filtering and locality-aware dissemination of information is a key requirement for assisted or autonomous driving as well as traditional assistance systems. In contrast to the work presented in this thesis, a distributed coordination of transitions might be required, as cellular connectivity cannot always be assumed. Moreover, locality-aware event dissemination mechanisms could benefit from additional locally available information, e. g., more accurate positioning or on-board sensors for speed and nearby vehicles.

Given the hybrid utilization of cellular and local ad hoc connectivity, the respective dissemination mechanisms could benefit from Software-defined wireless networks (SDWNs). Utilizing a SDWN, native multicasting of content to interested clients nearby could be utilized, for example, for clients associated to a Wi-Fi access point or an LTE cell. Similar to the popularity of SDN in fixed networks, SDWNs are expected to act as an enabler for more efficient dissemination mechanisms. This could include filter capabilities within the network itself, as already proposed for SDN [1, 2].

As briefly discussed in Chap. 1, we considered the reactive execution of transitions in the scope of this work. However, utilizing knowledge on application-specific behavioral patterns to execute transitions in a proactive fashion could further contribute to the seamlessness of transition execution. Proactive planning and execution of transitions is a core research topic in the second funding period of the Collaborative Research Centre "MAKI".

Lastly, transitions between filter schemes in a location-based publish/subscribe system can also act as an enabler for increased location privacy [3]. Instead of subscribing to an exact location (and updating that location frequently), a user might decide to switch to a less accurate filter scheme that does not require accurate location information. Such user-initiated mechanism transitions would enable users to select their preferred privacy vs. accuracy trade-off transparent to the application.

Our contributions to the realization and evaluation of mechanism transitions in publish/subscribe systems and their generalization in the SIMONSTRATOR.KOM platform constitute the foundation for further research in the aforementioned directions.

Acknowledgements This work has been funded as part of projects C2 and C3 by the German Research Foundation (DFG) within the Collaborative Research Center (CRC) 1053 "MAKI – Multi-Mechanisms Adaptation for the Future Internet".

References

1. Koldehofe B, Dürr F, Adnan Tariq M (2013) Event-based systems meet software-defined networking. In: Proceedings ACM international conference on distributed event-based systems (DEBS). ACM, pp 271–280
2. Koldehofe B, Dürr F, Adnan Tariq M, Rothermel K (2012) The power of software-defined networking: line-rate content-based routing using OpenFlow. In: Proceedings workshop on middleware for next generation internet computing. ACM, p 3
3. Onica E, Felber P, Mercier H, Rivi'ere E (2016) Confidentiality-preserving publish/subscribe: a survey. In: ACM computing surveys (CSUR) 49.2, p 27

Appendix A

A.1 Extended Study of Filter Schemes

This section provides additional evaluation results for the performance and cost characteristics of the individual filter schemes integrated into BYPASS.KOM (as discussed in Sect. 4.2.4). The evaluation is conducted in the same setup as described in Sect. 6.1 and results are reported for the DA mobility model and the AR workload, unless otherwise noted. In the following, we compare the performance and cost of individual filter schemes, discuss the influence of scheme-specific configuration parameters and, finally, evaluate the impact of reduced location accuracy.

A.1.1 Performance and Cost of Individual Filter Schemes

Figure A.1 shows the recall, precision, and traffic caused by the utilization of the respective filter scheme. We compare the characteristics under different workloads, with the workload models being discussed in detail in Sect. 6.1.2. The download traffic resulting from a scheme's utilization is directly tied to the precision of the scheme. A higher precision results in a reduced number of unnecessarily transmitted events. Consequently, the channel-based schemes GRID and EGRID cause significantly higher traffic than RADIAL and STE under all workload models. Compared to RADIAL as the scheme with the highest precision (0.98 on average), GRID with an average precision between 0.1 and 0.2 causes about four times higher traffic.

STE and EGRID both maintain a constant recall of 1.0. STE outperforms RADIAL in terms of recall, as it deals with mobility by intentionally extending the subscribed area in the direction of movement, as described in Sect. 4.2.4. This has an impact on the precision that is decreased to about 0.75. Still, the traffic caused by STE is only slightly higher than the traffic for RADIAL. This is due to the fact that the size of the extension shape defined in STE depends on a client's movement speed. If a client is

B. Richerzhagen, *Mechanism Transitions in Publish/Subscribe Systems*,
Springer Theses, https://doi.org/10.1007/978-3-319-92570-7

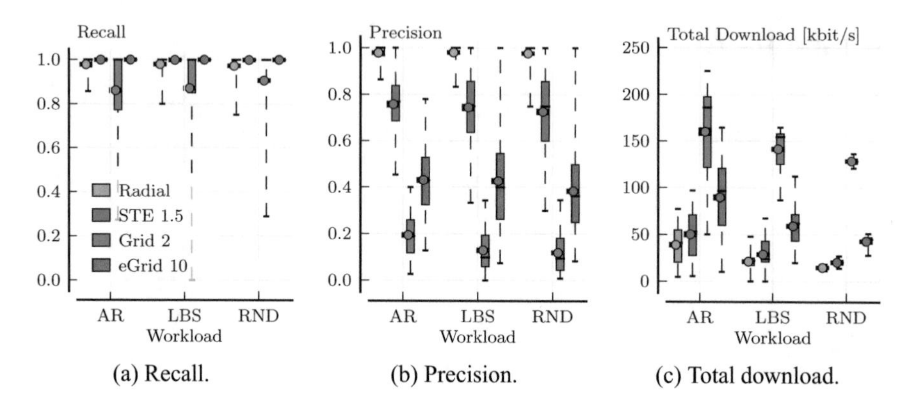

(a) Recall. (b) Precision. (c) Total download.

Fig. A.1 Performance of the individual filter schemes

not moving, STE behaves just like RADIAL, leading to high recall and precision and, consequently, low traffic.

Comparing the behavior for different workloads in Fig. A.1, we can observe that the results for the AR workload model are more skewed than the results reported for LBS and RND. This is caused by the heterogeneous load resulting from the AR model. In AR, load directly depends on the current location of clients and the resulting client density. The density is higher nearby attraction points compared to other areas, leading to the skewed results.

A.1.2 Impact of the Space-Time Envelope Size

The STE filter scheme can be configured with a parameter α that determines the size of the extension shape relative to the client's current movement speed. As introduced in Sect. 4.2.4, the length of the extension shape is calculated as $\alpha \cdot 60 \cdot |\mathbf{s}_u|$, with $|\mathbf{s}_u|$ being the client's movement speed in meters per second. In this section, we assess the performance and cost characteristics that result from different values of α. We vary α from 0.5 to 2.5 in 0.5 increments. Recall, precision, and the resulting traffic for different values of α are shown in Fig. A.2.

As expected, the recall of STE for $\alpha < 1$ approaches that of the RADIAL filter scheme. In the extreme case of $\alpha = 0$, one would observe the same filtering behavior as for RADIAL. The recall quickly saturates at 1.0 for STE with $\alpha \geq 1.5$. Higher values of α only lead to decreases precision and, consequently, increased traffic. We chose a default value of $\alpha = 1.5$ for our evaluations, given that this configuration leads to perfect recall of 1.0 at a reasonable decrease of precision to about 0.75.

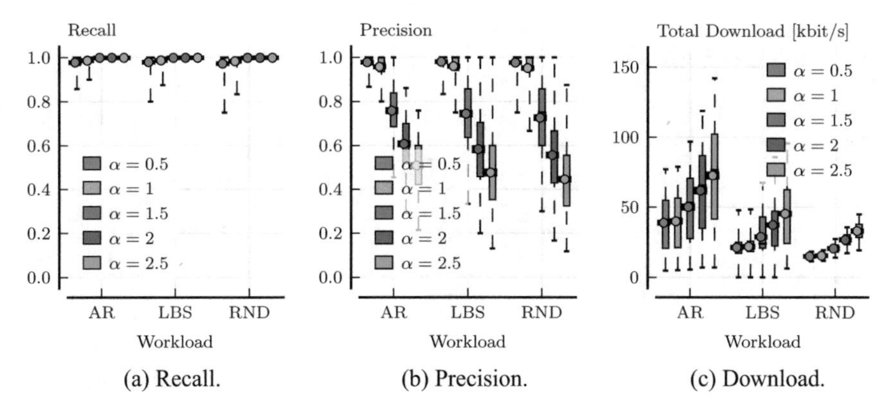

(a) Recall. (b) Precision. (c) Download.

Fig. A.2 Impact of α on the performance of STE

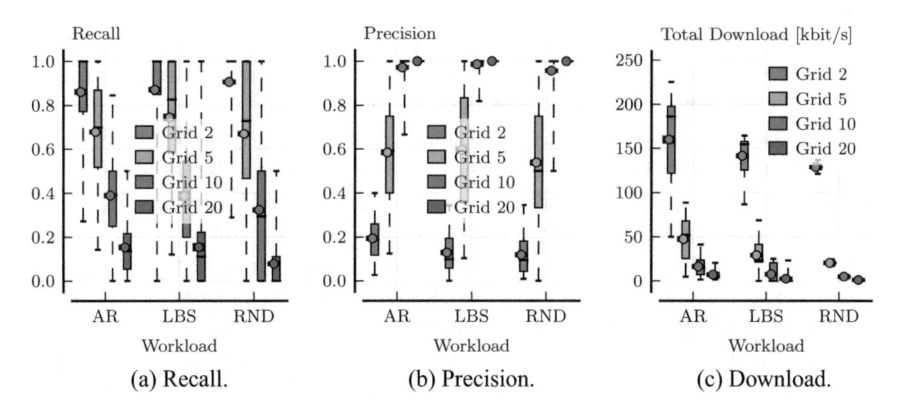

(a) Recall. (b) Precision. (c) Download.

Fig. A.3 Impact of different grid sizes on the performance of GRID

A.1.3 Impact of the Grid Factor on Channel-Based Filter Schemes

The performance of the grid-based filter schemes GRID and EGRID further depends on how well the chosen grid structure can reflect the subscriptions issued by clients' applications. In the following, we vary the *grid factor* introduced in Sect. 4.2.4. The larger the respective factor, the more cells are created. A grid factor of 2 leads to 4 channels, each covering a square area with an edge length of 650 m in our evaluation setup. With a grid factor of 20, we end up with 400 channels, each channel responsible for a square area with 65 m edge length. The radius of the AoI an application subscribes to is fixed to 125 m, as discussed in Sect. 6.1.

The results for GRID are reported in Fig. A.3. As clients only subscribe to one channel in GRID, the recall decreases quickly for larger grid factors. In order to achieve a reasonably high recall of above 0.85 for all workload models, GRID needs

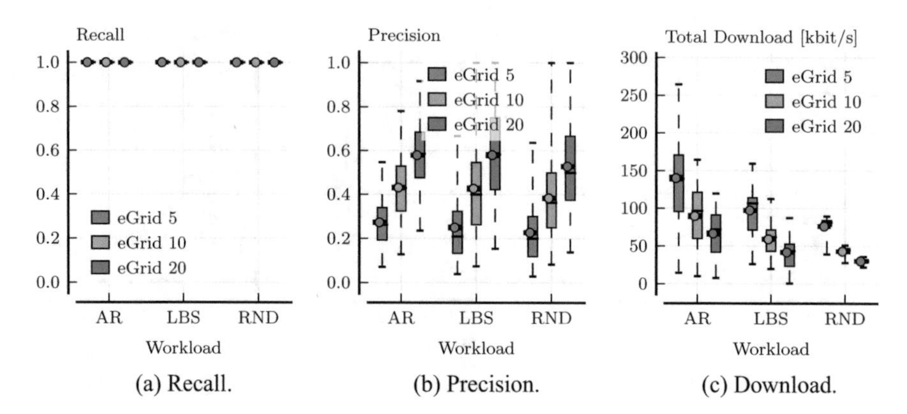

(a) Recall. (b) Precision. (c) Download.

Fig. A.4 Impact of different grid sizes on the performance of EGRID

to be configured with a grid factor of 2. However, this leads to a bad precision of below 0.2 on average. The traffic caused by GRID for a reasonably high recall exceeds that of all other filter schemes as already shown in Fig. A.1.

With EGRID, clients subscribe to all channels that fit within the MBR of their subscriptions, as discussed in Sect. 4.2.4. Consequently, the scheme is able to maintain a perfect recall of 1.0 under all workload models. The precision is directly determined by the size of the grid cells. For smaller cells (i. e., for a larger grid factor), the resulting MBR matches the actual AoI of a client more closely, leading to higher precision and, consequently, reduced traffic. At the same time, the coordination overhead due to more frequent channel reassignments increases (not shown here).

Given that EGRID relies on the MBR to assign channels to clients, the theoretical upper bound for its precision with respect to a circular AoI is $2/\pi \approx 0.64$. In the ideal case, the MBR corresponds to a square with edge length $2r$ (with r being the radius of the circular AoI) that is centered around the location of the client. In this case (and under the assumption of a sufficiently large number of events being published to equally distributed locations) the precision equals the ratio of both areas: $(4r^2)/(2\pi r^2) = 2/\pi \approx 0.64$. We observe higher values for the precision of a fraction of our publications in Fig. A.4, which is caused by the non-equal distribution of target locations in our workload models and the (comparably) small number of events.

A.1.4 Impact of Location Accuracy on Filter Schemes

In our evaluation, a new location is reported by the mobile OS every 5 s. Depending on the realization of the respective filter scheme, this updated location is sent to the broker (e. g., in RADIAL and STE) or used to check the current channel assignment locally (e. g., in GRID and EGRID). Within this evaluation, we vary the location update

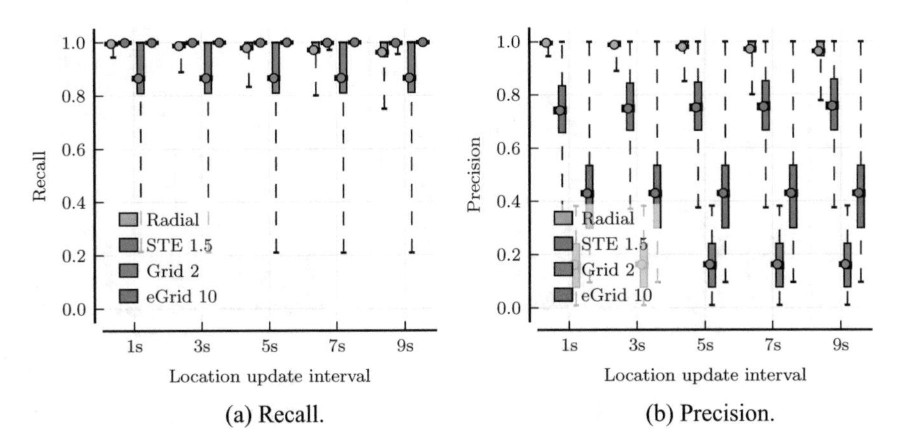

(a) Recall. (b) Precision.

Fig. A.5 Recall and precision for different location update intervals

interval to evaluate the robustness of the respective filter schemes to outdated location information.

Figure A.5 reports the recall and precision for different filter schemes depending on the location update interval. The channel-based filter schemes GRID and EGRID are robust to outdated location information given that the respective channels already cover an area exceeding the AoI of a client. For the interactive mobile social applications considered in our work, it is a reasonable assumption that updated location information is available every few seconds due to the utilization of GPS. In less interactive applications, longer location update intervals can lead to a performance degradation even for channel-based schemes.

RADIAL clearly suffers from reduced location accuracy, given that the average recall drops from around 1 for a location update interval of 1 s to about 0.95 for an interval of 9 s. Outdated location information further leads to a decrease in the precision of RADIAL. STE maintains an average recall of 1.0, given that the extension shape is designed to deal with continued client mobility if the movement direction remains similar. For a fraction of clients this assumption does not hold, leading to slightly decreased recall reported for the 2.5th percentile.

Figure A.6 reports the traffic caused by the filter schemes' context update protocols. While the absolute figures are are well below 0.5 kbit/s, their relative comparison allows us to assess how the respective filter schemes deal with location updates available on mobile clients. Given the large area covered by a channel in GRID with a grid factor of 2, the filter scheme's context update protocol is not affected by the considered location update intervals, as discussed previously. For channels covering a smaller area as considered in EGRID, less frequent location updates lead to a slight decrease in the number of reassignment requests sent to the broker. Consequently, the upload traffic caused by the context update protocol decreases slightly. Due to less assignment requests being issued by clients, the broker sends slightly less reas-

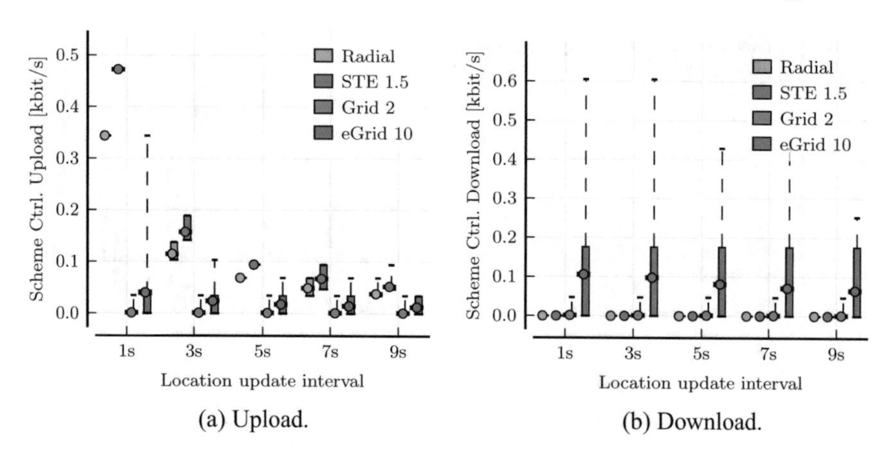

Fig. A.6 Client-side traffic caused by different location update intervals

signment messages to clients, leading to a reduction in the download traffic caused by the context update protocol.

Clients using RADIAL and STE report their current location whenever updated information is available. Consequently, the upload traffic decreases with longer location update intervals. In both schemes, the broker does not send any messages via the context update protocol, leading to a download traffic of 0 kbit/s. Messages issued by clients using STE are slightly larger than those issued by the RADIAL scheme. This is due to the fact that the current movement vector is included in the message to enable calculation of the extension shape at the broker, as discussed in Sect. 4.2.4.

A.2 Partial Transitions Between Filter Schemes

We compare the self-adaptive system MULTI against the execution of partial transitions between the ATTRACT and RADIAL filter schemes based on the location of attraction points. Augmenting the discussion in Sect. 6.2.4, we present the evaluation results for the MA mobility model in the following. We include the previously reported results for the DA mobility model for comparison. The overall characteristics in terms of recall, precision, and cellular traffic are shown in Fig. A.7.

For both mobility models (and the corresponding attraction points), including the subscription radius to extend the area covered by a channel in ATTRACT leads to a recall of approximately 0.99. The MA mobility model defines more attraction points than DA. This increases the area around attraction points where clients suffer from decreased recall if the radius of subscriptions is not taken into account, as shown in Fig. A.8a. Consequently, the average recall in the MA scenario reaches 0.88 compared to approximately 0.9 for DA, if radii are not considered. Additionally, the fraction of the simulated area that is covered by channels is higher than for the DA scenario, resulting in an overall decrease in precision.

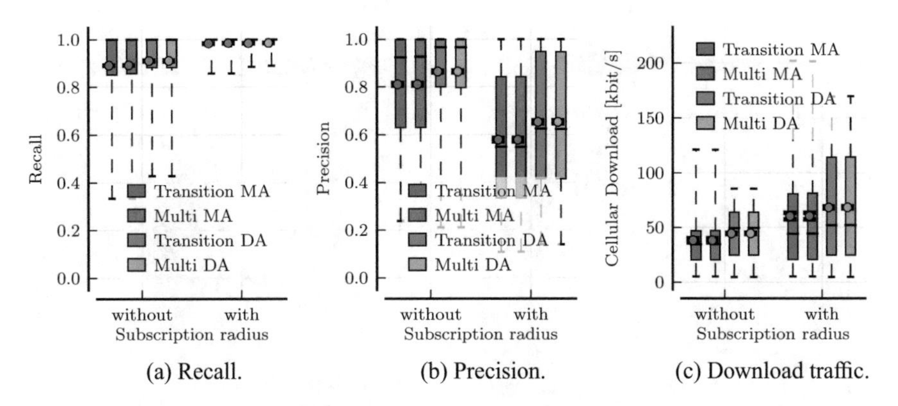

(a) Recall. (b) Precision. (c) Download traffic.

Fig. A.7 Performance and cost comparison

As previously reported for DA, including the radius of subscriptions when matching against the area defined by a channel in ATTRACT leads to a recall of approximately 1.0. This comes at the cost of reduced accuracy and, consequently, increased cellular download traffic, as reported in Fig. A.7c. The spatial distribution of recall and precision for the MA mobility model is shown in Fig. A.8 for completeness.

A.3 Extended Study of Dissemination Mechanisms

In this section, we present additional evaluation results that augment our evaluation of dissemination mechanisms presented in Sect. 6.3. We study the behavior of individual local dissemination mechanisms under different client densities. Further, we discuss the impact of the forwarding probability p on the performance of the GOSSIP mechanism. The setup of the respective evaluations remains as described in Sect. 6.1, unless otherwise noted.

A.3.1 Scalability of Local Dissemination Mechanisms

To assess the behavior of ad hoc dissemination mechanisms under different client densities, we utilize a ramp-based churn model as proposed in [1]. The number of active clients is increased linearly until it reaches a total of 300 clients after 25 minutes. We include the number of active clients in the plots presented in the following. All results are reported for the AR workload model. Consequently, the dissemination mechanisms are subject to geofencing, limiting forwarding of a local event to actual subscribers of the respective event, as discussed in Sect. 4.3.1.

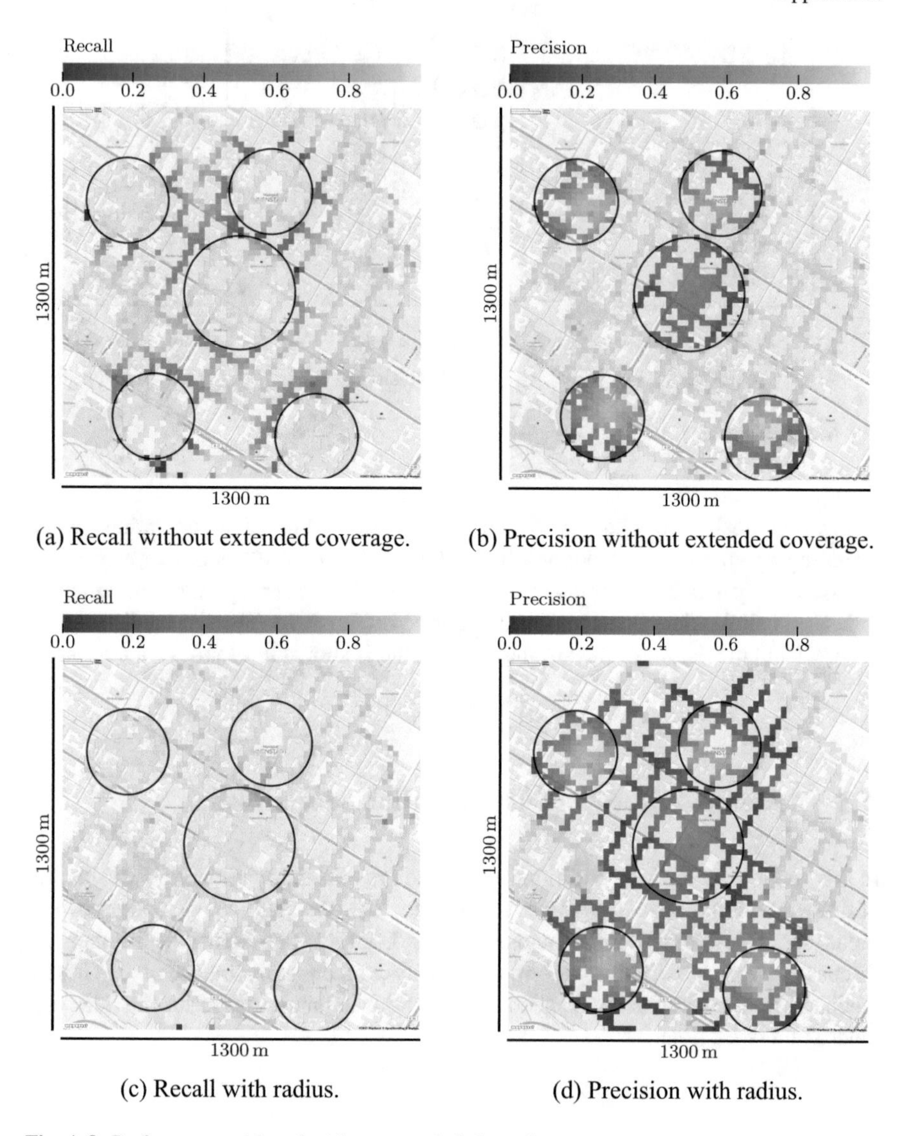

(a) Recall without extended coverage. (b) Precision without extended coverage.

(c) Recall with radius. (d) Precision with radius.

Fig. A.8 Performance with and without extended channel coverage

Figure A.9 shows the traffic characteristics of individual dissemination mechanisms with increasing number of clients. Additionally, the resulting mean dissemination delay for the distribution of events is shown. Regarding traffic, the aggressive mechanisms FLOODING and PLAN-B cause a significant increase for larger numbers of clients and, consequently, denser networks. GOSSIP and HYPERG exhibit roughly the same characteristics, causing only a third of the traffic of PLAN-B and approximately a fifth of the traffic caused by FLOODING. Traffic caused by the one-hop

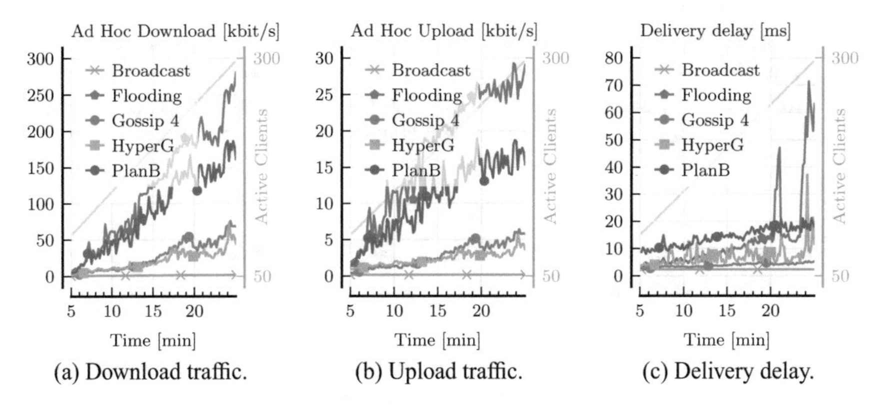

(a) Download traffic. (b) Upload traffic. (c) Delivery delay.

Fig. A.9 Behavior of dissemination schemes with increasing client density, AR workload

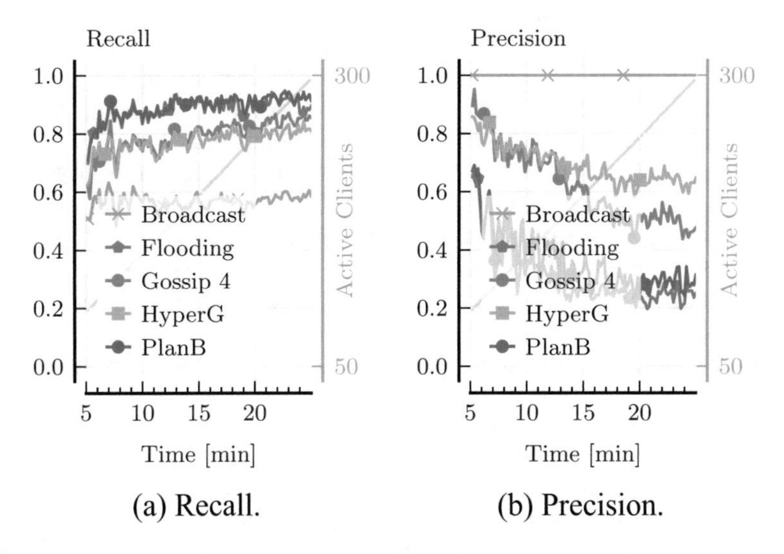

(a) Recall. (b) Precision.

Fig. A.10 Behavior of dissemination schemes with increasing client density, AR workload

dissemination mechanism BROADCAST is negligible in comparison. The delivery delay remains below 20 ms for all mechanisms, with the exception of FLOODING. FLOODING overloads the network, leading to increasing queuing times and, consequently, an increased delivery delay. For HYPERG, a slight increase in delivery delay can be observed towards the end of the simulated time. This is most likely caused by the retransmission mechanism that is used in HYPERG to deal with intermittent network partitions.

The recall of the dissemination schemes saturates quickly at approximately 0.9 for PLAN-B and FLOODING, as reported in Fig. A.10. The less aggressive schemes GOSSIP and HYPERG reach a recall of between 0.8 and 0.9, depending on the client density. With the AR workload about 60 % of subscribers are reached by single-hop broadcasting. The precision of all multi-hop dissemination schemes decreases with increasing client density. Single-hop dissemination with BROADCAST is not affected, as the transmission radius resulting from the network model configuration discussed in Sect. 6.1.3 is smaller than the radius of a client's AoI.

A.3.2 Impact of the Forwarding Probability on the Performance of Gossiping

We rely on GOSSIP with a default forwarding probability $p = 0.4$ for our evaluation of coexisting mechanisms and hybrid dissemination mechanisms presented in Chap. 6. In this section, we briefly discuss the impact of the configuration parameter p on the performance and cost of the GOSSIP dissemination mechanism. We consider the impact of client density as well as different workload models.

Figure A.11 reports the traffic resulting from different values of p. For $p = 1.0$, the mechanism behaves like FLOODING, as all messages are forwarded once by each client. As expected, lower values of p reduce the traffic significantly.

The resulting recall depends on the mobility model and the workload characteristics, as reported in Fig. A.11. For the AR workload, where events only need to be transmitted to nearby clients, a recall of 0.8 is achieved for $p = 0.4$. For larger values of p, the recall increases slightly, reaching 0.9 as previously reported for FLOODING. However, increasing the recall from 0.8 to 0.87 by changing p from 0.4 to 0.7 leads to a three times increase in traffic.

For other workloads, the impact of increasing p is more significant given that events need to traverse a potentially large amount of clients to reach their destination. Even with $p = 1.0$, we only achieve a recall of 0.4 for the RND and LBS workload models. This behavior is largely independent from the utilized mobility model, as reported in Fig. A.11d for the RND workload. With more messages being forwarded by clients, the overall delivery delay of events increases slightly, as shown in Fig. A.11e. GOSSIP remains stable for $p \leq 0.7$, while overload situations occur for $p = 1.0$ as reported previously.

A.3.3 Impact of the Mobility Model on Hybrid Dissemination

We discussed hybrid event dissemination modes in Sect. 6.3.2, showing the achieved performance and the resulting offloading benefits. However, we limited the discussion to the DA mobility model for brevity. In this section, we discuss the impact of different mobility models on the hybrid dissemination of events, further extending the discussion in Sect. 6.3.2.

Figure A.12 shows the performance of our hybrid modes compared to CLOUD and AD HOC. The hybrid dissemination modes AUGMENT and GATEWAY rely on the cloud

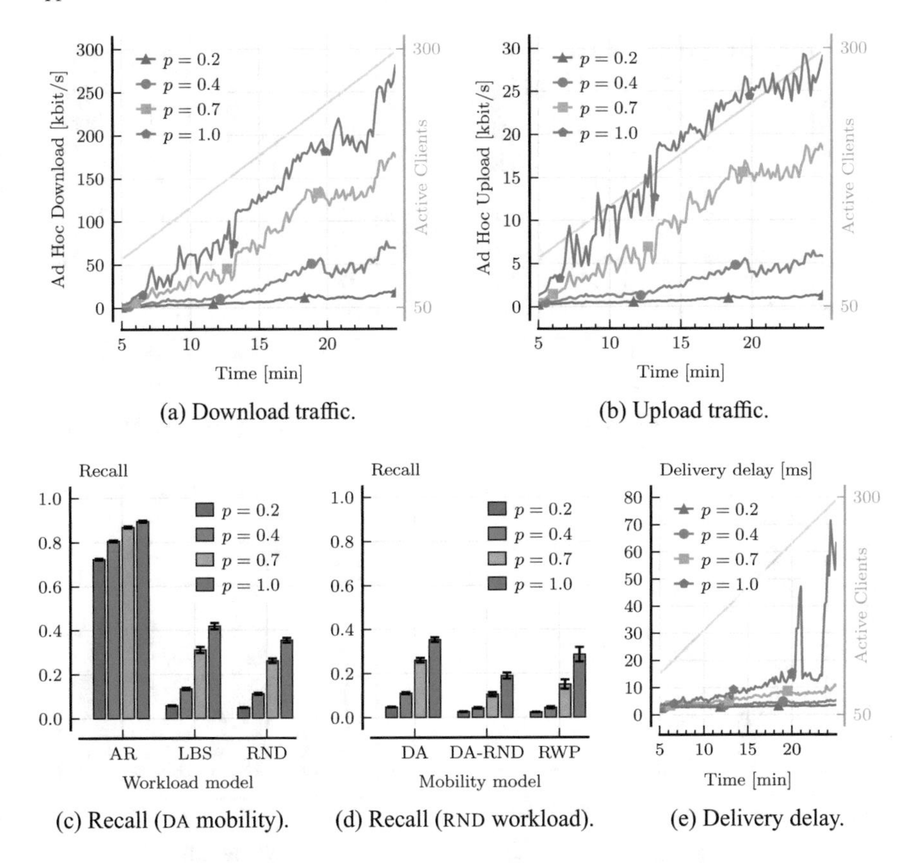

Fig. A.11 Impact of the forwarding probability p on the characteristics of GOSSIP

to ensure delivery of events, resulting in a constantly high recall of approximately 0.99. OFFLOAD suffers from the low recall of pure ad hoc event dissemination, leading to the respective recall of between 0.7 and 0.85 on average. The precision follows the same trend regardless of the mobility model.

However, the mobility model has an impact on the offloading effect achievable with the utilization of gateways. The offloading effect of the hybrid mode GATEWAY decreases if client mobility is no longer influenced by strong attraction points. Consequently, comparing DA and DA- RND, the offloading effect is reduced from approximately 75 % to 60 %. The same trend can be observed for the MA and MA-RND mobility models. For the LBS workload, the effect is less severe but follows the same trend.

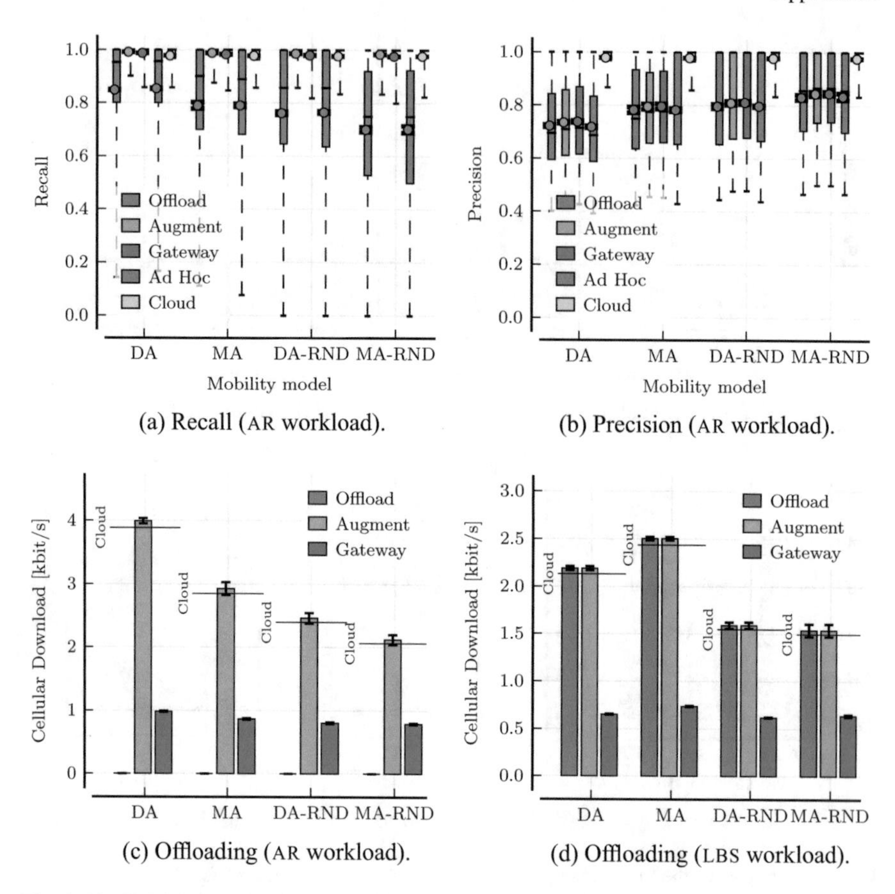

(a) Recall (AR workload).

(b) Precision (AR workload).

(c) Offloading (AR workload).

(d) Offloading (LBS workload).

Fig. A.12 Hybrid dissemination under different mobility models (AR workload)

Reference

1. Groß C, Richerzhagen B, Lehn M (2013) Structured search overlays. Benchmarking peer-to-peer systems. Springer, Berlin Heidelberg, pp 49–67

Author Biography

Dr. Björn Richerzhagen is currently a Postdoctoral Researcher at the Multimedia Communications Lab (KOM) at the Technische Universität Darmstadt, Germany, heading the research group Distributed Sensing Systems (DSS).
This book is the result of Björn's Ph.D. thesis in Electrical Engineering and Information Technology with a focus on Computer Engineering, conducted in the Collaborative Research Centre (CRC) 1053 "MAKI – Multi-Mechanisms Adaptation for the Future Internet" approved by the German Research Foundation (DFG). He was awarded his doctoral degree (Dr.-Ing.) in July 2017 from the Technische Universität Darmstadt. The thesis was awarded best dissertation in the Department of Electrical Engineering and Information Technology by the Vereinigung von Freunden der Technischen Universität zu Darmstadt e.V. (Association of Friends of the Technische Universität Darmstadt) for its outstanding scientific contributions. Björn received his Master of Science in 2012 and his Bachelor of Science in 2011 from the Technische Universität Darmstadt.
Björn has published his results in over 30 journal articles and peer-reviewed conference papers. His research interests include distributed event-based systems, most notably peer-to-peer systems and sensor networks, with a focus on their utilization in highly dynamic mobile environments.

B. Richerzhagen, *Mechanism Transitions in Publish/Subscribe Systems*,
Springer Theses, https://doi.org/10.1007/978-3-319-92570-7

Printed in the United States
By Bookmasters